AF840210

Philosophy of Science for Theologians
An Introduction

Contributions to Philosophical Theology

Edited by Gijsbert van den Brink, Vincent Brümmer and Marcel Sarot

Advisory Board:
David Brown
Paul Helm
Eberhard Herrmann
Werner Jeanrond
Christoph Schwöbel
Santiago Sia
Alan Torrance
Nicholas Wolterstorff

Vol. 12

PETER LANG
Frankfurt am Main · Berlin · Bern · Bruxelles · New York · Oxford · Wien

Gijsbert van den Brink

Philosophy of Science for Theologians

An Introduction

PETER LANG
Internationaler Verlag der Wissenschaften

Bibliographic Information published by the Deutsche Nationalbibliothek
The Deutsche Nationalbibliothek lists this publication in the Deutsche Nationalbibliografie; detailed bibliographic data is available in the internet at <http://www.d-nb.de>.

This publication was made possible thanks to financial support from the Dutch Organisation for Scientific Research NWO ('Nederlandse Organisatie voor Wetenschappelijk Onderzoek') and from the Protestant Theological University (PThU).

An earlier edition of this book was published in Dutch under the title: "Een publieke zaak. Theologie tussen geloof en wetenschap".
© Copyright 2004 by Boekencentrum Uitgevers, Zoetermeer / The Netherlands.
This edition has been translated into English by Chris Joby, Durham.

ISSN 1433-643X
ISBN 978-3-631-56951-1

© Peter Lang GmbH
Internationaler Verlag der Wissenschaften
Frankfurt am Main 2009
All rights reserved.

All parts of this publication are protected by copyright. Any utilisation outside the strict limits of the copyright law, without the permission of the publisher, is forbidden and liable to prosecution. This applies in particular to reproductions, translations, microfilming, and storage and processing in electronic retrieval systems.

www.peterlang.de

To Séan

Preface

The aim and content of this study can appropriately be summarized in a single sentence: This book tells the exciting story of the philosophy of science from its inception in the aftermath of the first World War to its current stage, and relates this story to theology. In doing so, it fills a remarkable gap in the literature. There are of course numerous studies – among which some recent ones – that describe the history of the philosophy of science, either or not with a special eye for a particular discipline or group of disciplines (natural science, the social sciences, the humanities et cetera). But none of these books focuses on theology, or even pays a considerable amount of attention to it. Conversely, there is a lot of discussion about the status of theology among theologians and academic policy-makers in the Western world. Does the age-old discipline of theology, with which long ago the university started, still deserve a place at our largely secularized universities? Or should faculties of theology be reshaped into more 'neutral' departments of religious studies, which in turn might most efficiently be relegated to some larger faculty? The enormous resurgence of religious issues in often heated discussions both in politics and in society since the beginning of the 21st century gave a new urgency to the question of the academic treatment of religion(s). Is it still adequate to allow for the academic study of religion only in the distanced and matter-of-fact way that is characteristic for religious studies, without people's own views of life being brought into play and confronted with each other? Can questions on the meaning, value and truth of religious convictions be kept out of the door so easily? There is a wide debate on topics like these – but seldom this debate on the status of theology and religious studies as academic disciplines is conducted in a way that is informed by the state of the art in the philosophy of science.

So this book tries to remedy this situation. It was written for everyone who has an interest in the relation between (Christian) theology and the sciences, but with a special focus on students specializing in theology and/or religious studies. Its origin lies in lectures on the theory of science which I gave at the (by then) theological faculty of Utrecht (Netherlands), and I have tried to maintain the traces of this educational background. For it seems to me that a basic knowledge of the philosophy of science as well as a training in applying its history to questions concerning the academic status of theology should be part and parcel of a proper theological education. Hopefully, this book can be useful in this connection, either as a course book or in some other way. Although the topics it deals with are complex and the use of technical language cannot be avoided in explaining them, I have tried to write as clear and accessible as possible.

Philosophy of Science for Theologians goes back to a Dutch precursor, which appeared under the title 'A Public Concern. Theology Between Faith and Science' (*Een publieke zaak. Theologie tussen geloof en wetenschap*; Boekencentrum, Zoetermeer 2004). I am tremendously grateful to Dr. Chris Joby from the University of Durham for translating this work into English, following on from a pilot translation of one chapter by Dr. Gerrit Brand from South-Africa, to whom I am indebted as well. I have compared the translation sentence by sentence to the Dutch original, and I was impressed both by the difficulty of the job and by Dr. Joby's carefulness. Meanwhile, I adapted the English text to a new and wider audience, replacing many references to literature which is only available in Dutch by English sources. I have not, however, deleted all references to Dutch publications. After all, the story of this book can only be told by discussing the work of British, American, German, Hungarian, Austrian, French, South-African and many other authors, so it would be unfair to systematically exclude the Dutch for the sole reason that their work is not always available in English. I hope for the reader's understanding at this point – perhaps he or she may even appreciate the opportunity to get a glimpse of what is going on in Dutch theology and philosophy.

Going through the English version also gave me the opportunity to incorporate and sometimes briefly discuss a number of more recent publications that appeared after the publication of *Een publieke zaak.* More generally, at some places I felt challenged to express myself more clearly than I had previously done, adding some clarifications or changing the wording of a particular paragraph. In some of these cases I could profit from constructive criticism that was uttered in reviews of the Dutch original. Further, I dropped two complete sections (one on the Dutch theologian H.M. Kuitert and the other on the Dutch academic scene in relation to theology) which were so closely bound up with the Dutch situation that they might hardly appeal to a wider audience. And I decided to include some new material instead – the most important of which is to be found in section 5.2. Here, it occurred to me that I had been doing insufficient justice to Karl Barth's radical view of theology as a church discipline which should not be measured by (or attempt to satisfy) general standards of scientific inquiry. Although in the end I still disagree with (at least the implications of) this view, I felt obliged to do more justice to it, given the fact that it continues to be a powerful alternative to the approach adopted in this book. It may be clear that, given the number of changes I made in Dr. Joby's text, the book's writer and not its translator is to be blamed for any infelicities in the English that have resulted from this process. Since this English edition deviates in quite a number of respects from the Dutch original, I decided that it might be misleading to use a literal translation from the Dutch as its title. Therefore, I chose a new title that expresses the book's intention more straightforwardly.

Finally, a lot of colleagues have contributed to the development of this book, either by carefully reading and reviewing parts of it or by talking with me about its themes. In this connection, I would like to thank H.J. Adriaanse, Gerrit Brand, Willem B. Drees, Jan Hoogland, Barend Kamphuis, Gerrit de Kruijf, Remko Muis, Arie de Reuver, Marcel Sarot, Taede Smedes and Wessel Stoker for their critical comments on earlier drafts of this book. Many thanks also to Gerton van der Wilt, who helped me to prepare the index. I am grateful to Prof. Wentzel van Huyssteen (Princeton Theological Seminary) for stimulating me to prepare an English version of the Dutch original. Having dedicated earlier books to our two older children, I dedicate this work to our four year old foster son Séan.

VU University Amsterdam, August 2009 *Gijsbert van den Brink*

Table of Contents

1. Science, religion and society

1.1 An outdated discussion?

It might be thought that the debate about faith and science has had its time. I have been in the situation where someone I was speaking to frowned at me, looking somewhat surprised, when I told him that as a theologian I was fairly actively involved in this debate. According to him, I was running the risk of tilting at windmills. In our post-modern times, he thought, faith has hardly anything to fear any more from the pretensions of 'science'. Indeed, in the debate in society at large, religion and faith do not need to justify themselves any more in the face of an allegedly unassailable Reason. Quite the opposite in fact: a strong tendency towards the irrational can be observed. For many, it is now not what reason prescribes that is determinative, but what feeling experiences. Miracles are no longer 'forbidden' any more, and in the media today miracle stories even lead to high viewing figures. The belief in a hereafter is increasing rather than decreasing. And the theory of secularization – the theory that religion and faith would disappear of their own accord as the process of modernization took hold – meanwhile seems to have been abandoned everywhere. Characteristically, one of the most influential sociologists who introduced this theory in the 1960s, Peter Berger, has subsequently acknowledged that this was the biggest mistake of his career.[1] So, religion has proved to be highly durable, and even in the Western world at the beginning of the twenty-first century, its presence is undiminished.

We can ask ourselves whether the post-modern era that we are currently in does not in fact bring with it a set of challenges to religious belief that are completely different from those that are presented by science. For example, that of relativism. Does not the Christian faith have much more to fear from a relativist and pluralist social environment, in which it does not really matter what you believe – as long as it makes you feel good? The question of what may ultimately be called truth is no

1 Peter L. Berger, 'Protestantism and the Quest for Certainty', *The Christian Century* 115 (1998), 782-796. 'In the course of my career as a sociologist of religion I made one big mistake and had one big insight (arguably not such a bad record). The big mistake, which I shared with almost everyone who worked in this areas in the 1950s and '60s, was to believe that modernity necessarily leads to a decline in religion. The big insight was that pluralism undermines the taken-for-grantedness of beliefs and values' (782).

longer of concern. Now it is just tastes that differ. The view that different religions and religio-philosophical movements claim *essentially* different things, and that it really does make a difference whether you are a Muslim, a Christian or an atheist, is one that is liable to wear out quickly. It is just this all-pervasive feeling that it does not really matter in the end whether you believe and what you believe (a feeling that the Dutch theologian, Anton Houtepen, some years ago aptly described with the term *agnosm*[2]), that could in fact be the greatest modern-day challenge to the Christian faith. So, the point is no longer the same as in the nineteenth century, when all sorts of public *arguments* against faith were trotted out, the merits of which you could then debate, but rather that faith is as it were undermined from within. Is it not this that theologians should be reflecting on?

This last point is doubtless true. However, contemporary developments in philosophy and religion are too complex to be reduced to one single process. In our West European culture, we have to deal with a number of 'atheizing factors', i.e. factors that contribute to the fragmentation of the classical belief in a personal, active God, which has been characteristic of Christian tradition down the ages. Some of these factors are without doubt related to the shift from a modern to a postmodern era: pluralism, relativism and rampant individualism take their toll. It is rather more difficult, however, to trace other factors back to this shift. I am thinking of our increased sensitivity to the seriousness of *evil* in the world. The whole problem of wide-scale and world-wide evil and suffering such as that which we see and hear of in the media every day is of course not the least of these 'atheizing factors'. Many have abandoned God and faith with the question which has almost become a cliché: 'If there is a God, why does that God let evil happen?' Sometimes this question is asked with the best of intentions in the context of an intense existential struggle. At other times it seems like a welcome excuse to be able to bid farewell to what in any case someone had already had enough of – the question is sometimes posed almost casually, with a simple shrug of the shoulders. Other people, it seems, are impressed by developments in philosophy, which have apparently demonstrated that it is no longer sensible to believe in the personally active God of the Bible. We need think only of 'great names' such as those of Nietzsche, Heidegger, Habermas and many others (including post-modern philosophers).[3]

2 A. Houtepen, *God: An Open Question,* New York 2002. Others use the word 'apatheism' to describe the same attitude, e.g. Charles Mathewes, 'Political Theology After the End of History', unpublished paper presented at the Annual Conference of the SST, Durham, April 2008, 8.

3 I have previously attempted to evaluate the importance and validity of the two atheizing factors mentioned last (the problem of evil and the developments of philosophy) resp. in *Almighty God. The Doctrine of Divine Omnipotence*, Kampen 1996[2] (1993) and in *Oriëntatie in de filosofie*, Zoetermeer 2000.

In the midst of all these sorts of atheizing factors, *science* too still plays a role which should not be underestimated. It seems in fact that whilst in days gone by only the top echelons of society saw a conflict between science and belief, this idea did not filter through into the rest of society until the second half of the last century. The story behind the bestseller which the Delft physicist A. van den Beukel wrote, *More things in heaven and earth*,[4] is worth recalling in this regard. Van den Beukel came to write this book about the relation between faith and science after his brother, who was a pastor, had complained about the fact that aspiring catechumens regularly failed to turn up, using 'science' as their excuse. 'Science has now for a long time shown us that God does not exist, so why should we still be involved with the church?' is a line of reasoning which he often heard. Theoretical physicist Van den Beukel considered this reasoning so absurd, that he felt moved to make a comprehensive response.

That is not to say that anyone who thinks more deeply or who has more knowledge of these things would not see a conflict between faith and science. Quite the contrary. Today, there are still eminent scholars who subscribe to the same way of thinking as the catechumens we have just discussed. Most well-known perhaps is the work of scholars such as Richard Dawkins, Edward O. Wilson and Daniel Dennett, who argue that contemporary evolutionary theory cannot only fully explain the phenomenon of religion, but in doing so also makes religion superfluous, showing that God is a 'delusion' (Dawkins). In the Netherlands, H. Philipse and H. J. Adriaanse, amongst others – also stress that the progress of science has seriously affected the plausibility of belief in God, if not completely undermined it.[5] This in fact concerns any and every belief in a personal, active God, including therefore – and this is what we are particularly interested in here in this book – classical Christian belief.

To be able to situate this sort of statement (which is echoed by many others in this field), we must recognize that it is not entirely correct to say that our era is one that is simply 'post-modern'. In reality the spiritual climate of our times shows traces of modernity as well as of post-modernity. The lines of influence which can be traced back to the eighteenth century Enlightenment have definitely not yet been broken. In some respects (think of individualism) they have only

4 A. van de Beukel, *More Things in Heaven and Earth: God and the Scientists*, London 1991.

5 Cf. e.g. H.J. Adriaanse, *Vom Christentum Aus. Aufsätze und Vorträge zur Religionsphilosophie*, Kampen 1995, 81: 'Zehnmahl will ich gesagt haben, dass Wissenschaft dem christlichen Glauben Abbruch tut'; H. Philipse, 'The Incompatibility of Science and Religion: An Argument for Atheism', in: William Desmond, John Steffen, and Koen Decoster (eds.), *Beyond Conflict and Reduction. Between Philosophy, Science, and Religion*, Leuven 2001, 117-134.

been *strengthened* by post-modernity. In other respects – the glorification of reason, the idea of progress – they have been seriously damaged by it. But even so, modernity remains a factor whose importance should not be underestimated. It is certainly true that 'faith in science' is no longer so unremitting as it was in the days of Rudolf Bultmann. Bultmann was able to argue that the invention of radio (!) – welcomed by the Dutch theologian and statesman, Abraham Kuyper, as an *aid* to faith: was not wireless technology, he asked, a practical illustration of how *prayer* works?[6] – makes it impossible for us to take the depictions of faith in the New Testament seriously.[7] Nowadays, this is no longer repeated very often anymore But the *effect* of contemporary science is still to a considerable degree atheizing in character.

For this reason, we are definitely not fighting a rearguard action, if we occupy ourselves in this study with the relationship between faith, theology and science. Rather, we are having a debate that still remains completely open. This is clear from the fact that on an international scale we find a huge, and ever increasing interest in this theme. We can even say without too much exaggeration that *Religion and Science* or *Science and Theology* is actively developing into a separate field of study, with its own journals, special professorial chairs, international academic societies etc.[8] And it would not be the first time that a new, more or less autonomous, area of research and teaching has split itself off from the catch-all discipline we call philosophy. In this new area, we can hear both voices that plead for more co-operation and interaction between faith and science, and one that argue that each of them is different, and has its own function and meaning in our lives, and that the boundaries between them should therefore be carefully maintained.[9]

6 A. Kuyper, *To Be Near unto God*, Vancouver BC 2005 (original edition Grand Rapids 1918), 50 ('This now comes to help our weak faith'), 341; cf. Alvin Plantinga, *Warranted Christian Belief*, Oxford 2000, 404.

7 'We cannot use electric lights and radios and, in the event of illness, avail ourselves of modern medical and clinical means and at the same time believe in the spirit and wonder world of the New Testament'. Rudolf Bultmann, *The New Testament and Mythology, and Other Basic Writings*, Philadelphia 1984, 4.

8 An example of the first is the journal started in 2003, *Theology and Science.* An example of the second is the appointment of J. Wentzel van Huyssteen as 'Professor of Theology and Science' at Princeton Theological Seminary, and two examples of the third are the *European Society for the Study of Science and Theology* (ESSSAT), founded in 1988, and the *International Society for Science and Religion*, founded in Granada (Spain) in 2002.

9 A recent example of the first is Alan G. Padgett, *Science and the Study of God. A Mutuality Model for Theology and Science*, Grand Rapids 2003. A recent example of the second Taede A. Smedes, *Chaos, Complexity, and God. Divine Action and Scientism*, Leuven 2004.

To put our cards on the table straight away: in this book we engage with the discussion to show that the supposed atheizing effect of science does not accord with its real nature. Those who suggest the opposite – i.e. that atheism or agnosticism do accord most closely with the nature or results of science – in my opinion overplay their hand, and in the background there always turns out to be an ideological prejudice at work. In this regard, we rightly talk of *scientism*, where the ending -ism indicates that we dealing with a non-rational attempt to overestimate and absolutize the goals of (natural) science.[10]

Apart from the categories 'faith' and 'science', there is a third category that plays a role in this context, namely one that we shall denote with the term 'theology'. Actually, this concept is the most important focus of our discussions in this book. Traditionally, theology has occupied a place between the two poles of faith and science. But where exactly? Can it be understood completely as a science? Or should we perhaps totally reject this idea? Should theology be content with the articulation of certain (irrational) types of personal faith that some people hold to? In this book we will argue that the opposite is the case, namely that theology is a completely public matter. In theology what is under discussion is not only a pleasant pastime for a group of insiders, but something that concerns everyone who thinks more deeply about life and wants to think seriously about life's big questions.

Before we go on to discuss the layout of our study, we shall first of all look somewhat more closely at what we mean when we talk of 'science' in the rest of the book. In our technologically highly-developed society we have in fact become so accustomed to the public and private use of all sorts of applications of science, that we hardly notice any more how deep the scientific-technological revolution of the last few centuries has become enmeshed in our experience of and dealings with reality. In countless ways, our lives have been unimaginably changed by it in comparison with those of previous generations. There can be no doubt that to a large extent we can view these changes as changes for the better, and as it happens almost without exception we do view them as such. That is why, first of all, we briefly want to remind ourselves of the enormous blessings that the developments in science and technology have brought with them so far. This will prevent us – whether or not in the name of faith – from merely adopting a hypercritical attitude towards science and its results.

10 Compare the monograph of the Swedish philosopher of religion Mikael Stenmark, *Scientism. Science, Ethics and Religion*, Aldershot 2001. Stenmark shows that modern scientists such as Richard Dawkins and Edward O. Wilson are insufficiently aware of the fact that their arguments presuppose certain non-scientific, ideological beliefs (in this case, of a materialist nature).

1.2 The blessing of science

Science, more than any other phenomenon, has radically altered the appearance of the world in the past few centuries. It is difficult to overestimate the influence that the results of scientific research have on modern life, although in reality we always tend to underestimate this influence. This happens because we have rather short memories. As a result, we tend to consider all sorts of things as self-evident, that are in fact far from being so. Through a thousand and one applications of technology, science has fundamentally changed our worldview since the seventeenth century in quite a number of fields.[11] Think for example of the greatly increased possibilities of medical care, the mechanization (and the present wide-scale digitalisation) of the production process of economic goods, of the manner in which and the speed with which we move from place to place, of the rise of new media with their rapid provision of information, of the continually progressive development of communications technology, and of many other things.

Things have moved on so much that it is almost impossible now to imagine how they would look without all the technological innovations of recent centuries. I remember how, during my days as a student, a professor wanted to convince us of the enormous impact of the scientific revolution by talking for half a lecture about the tap. This was so surprising because it must be a long time since we last thought of such simple and everyday things as taps when using terms such as 'science' and 'technology' as they are generally understood. Whilst personal computers, airplanes and satellites still fire the imagination, it has been a long time since the tap has done the same. But what unimaginable changes did the laying of a network of water pipes bring to the daily rhythm of the lives of people in the West! For the first time people could make quantities of water of, in theory, unlimited proportions, flow easily and conveniently into their own kitchens, or into any other room they wanted. And they could do all this by means of one simple movement of the hand to turn on a tap. It must have meant an enormous saving of time no longer to have to take water home from a well or pump one bucket after another. So the invention of the tap (and the transporting of water through iron pipes) at the end of the eighteenth century is, all things considered, a clear example of the progress which went hand-in-hand with the development of science and technology.[12]

11 See on this process e.g. the classic work of E.J. Dijksterhuis, *The Mechanization of the World Picture. Pythagoras to Newton*, new edition, Princeton 1986 (original English edition 1961).

12 At the end of the eighteenth century, the invention of the steam-engine and the use of iron pipes in fact made it possible to transport water to the places where it would be used in people's houses. In the Netherlands, for example, the first water company

Indeed, we should not shrink from using that word 'progress'. An enormous amount of criticism has been directed at the so-called 'idea of progress', some of which is no doubt justified, and later on we ourselves shall to a large extent also critique and qualify this type of thinking – but that does not take away the fact that the development of science has in fact brought mankind enormous improvements in a large number of fields. The fact that we all agree about this can be seen from the massive extent to which we make use of most technological developments.

We can have very ambivalent feelings about the developments of science and we can be rightly apprehensive about the risks and the negative aspects associated with it. We may even think that the whole intellectual climate within which science and technology could and can continue to flourish should be treated critically, because it presupposes a humanity which grossly over-estimates itself, as if we should and could shape our lives by ourselves as much as possible. The whole idea of a society that can simply be stage-managed can be whole-heartedly rejected. But all this does not take away the fact that we shall continue to make use of water pipes, roads, medicines, modern means of transport and communications, electricity, computers, etc. etc. For as long as we have these at our disposal, we shall in fact almost without exception make use of them. In any case, I know few people who are, for example, as a matter of principle, against travelling in cars (although there are somewhat more people who as a matter of principle do not want to learn to drive one), and no-one at all who categorically rejects the use of the tap. This last is equally true for countless other technological applications that have been produced as a result of scientific research: no right-minded person refuses to make use of them. And from this simple fact, it can be concluded that in practice we consider these sorts of 'modernizations' as 'improvements' (for why else would we make use of them?), and so, as progress.

We really want to emphasize this point here. In the course of this book, we shall take a rather critical view of all sorts of pretensions and positions that are presented as 'scientific'. What is more, we will comment critically on the aura of comprehensiveness associated with the scientific enterprise as such. We shall ask penetrating questions about what we see as the absolutization and over-estimation of science in large sections of modern Western society. However, all this should in no way simply be confused with a position that is hostile to science. We do encounter such a position here and there, on the one hand amongst strongly religious people, on the other in left-wing circles that are strongly critical of existing social structures. Sometimes each new development in or by the scien-

was founded in 1853 (it transported water from the dunes in North Holland to Amsterdam), and before 1900 all large towns had a good water system. After 1910, the countryside was also gradually provided with (regional) water systems.

ces seems to be the object of suspicion right from the outset. 'Science' is simply seen as the enemy of all that people cherish and hold dear. But such a position is wrong-headed. And not only that: it is also untenable – because and in so far as we ourselves frequently and in countless ways make use of the results of science. This is still the case, even if we limit ourselves to technology that has been in existence for a considerable time, or if we make use in general of modern technological innovations in as restrained and (from the point of view of environmental considerations) responsible a way as possible.

Here, the integrity of our lives is at issue. For we are guilty of double standards if on the one hand we reject all the claims of 'science', and on the other hand in practice welcome its results almost as eagerly as others do. Just as it is in a certain sense dishonest to do the latter in some areas in a rather uncritical manner, and then to relate super-sceptically from the outset to all scientific claims in other areas (e.g. where science involves itself with the sources of our faith). In this regard, Bert Loonstra is right when he argues that to be consistent we should go to live like the Amish 'who combine their rejection of modern science with the rejection of all the achievements that have arisen from scientific endeavour'. But, if we do not do that, we cannot simply reject everything that presents itself as science/scientific in one fell swoop, but instead in the first instance we ought to consider carefully its truth content.[13]

In short, there is no reason for us to begin by taking a sceptical or critical position in relation to the modern scientific enterprise and its applications – and anyone who suggests otherwise should check carefully whether their lifestyle accords with their position in this regard. Instead, we should be very thankful for the many possibilities that we can avail ourselves of as a result of scientific developments, possibilities which all previous generations had to do without. The developments in the sphere of medicine, for a start, provide us with sufficient occasion to be thankful. What a blessing it is that we can be helped in a professional and efficient manner when, for example, we have toothache (an ailment from which an enormous number of people have suffered in days gone by!), and what a blessing it is that people no longer have to die from diarrhoea, and that we have been able (for the time being) to eradicate a number of other infectious diseases, etc. Even psychological depression, that many were victim to in the past (including those around them) without knowing what exactly it was and where it came from, can nowadays sometimes be effectively combated with psychiatric drugs and other means at our disposal.

13 Bert Loonstra, *God schrijft geschiedenis. Disputaties over de Eeuwige*, Zoetermeer 2003, 19f.

Of course, people can immediately point to all sorts of dangerous and negative consequences that result from this same scientific development. Regarding medicine, we might think for example of the whole question of euthanasia, that comes up because doctors can carry on treating us almost forever. And as regards the natural sciences, we cannot of course ignore the production of weapons of mass destruction which they make possible. It is still far from impossible that sooner or later these weapons will fall into the wrong hands, and cause death and destruction on a huge scale across our planet. We could list yet more negative and life-threatening consequences of the development of the sciences, for example in the field of ecology and the environment. We should not close our eyes to these sorts of worrying consequences, in a blind glorification of science and technology.

However, in my opinion we must not be too quick in pointing out these negative aspects of technological developments to denigrate science as a whole. Let us rather first say, in an open-minded and unambiguous way, that it is a privilege to be able to live in a period in which we can make life more comfortable in so many ways thanks to science and its applications. To deny this positive side of things, or to play down the very real improvements in living conditions that we owe to developments in science, would be the height of ingratitude – if only towards our forefathers, who would have been happy to share in many (although no doubt not all!) of the results of the scientific revolution.[14]

It is for this reason that we begin by focussing on what we would unreservedly want to call the blessing of science. This blessing can be seen most clearly in the realm of medical science, although it can of course also be seen in other areas. There is therefore no reason whatsoever to be afraid of science, or to oppose the practitioners of science from the outset with scepticism or suspicion. On the contrary, we have had a particularly large amount of things to be thankful for in the last few centuries as a result of their activities. In the course of this study, we shall briefly and critically comment on all sorts of 'scientific belief', i.e. on the sometimes boundless over-estimation of the possibilities and the equally serious under-estimation of the limits of science. However, this generous and unambiguous recognition of the manifold and remarkable positive effects of science should be a point of departure that no-one can disagree with.

14 With the term 'scientific revolution', we mean the liberation of the natural sciences from philosophical frameworks, which began in the 16th century and reached a high point in the work of Isaac Newton. See e.g. A. Rupert Hall, *The Revolution in Science, 1500-1750*, London 1983; and the sophisticated treatment by Peter Dear, *Revolutionizing the Sciences. European Knowledge and Its Ambitions, 1500-1700*, Princeton 2001.

1.3 Structure

In this study, we shall proceed as follows. We shall begin with a somewhat more precise description of the phenomenon of science, or to put it more accurately: of the extremely interesting development that the philosophical reflection on it has undergone in the past century. This philosophical reflection has even developed into a separate discipline within the larger subject of philosophy, namely the philosophy of science. Anyone who wants to know how exactly they should assess the phenomenon of science according to its type and results, possibilities and limits (and this seems to be something we cannot avoid if we want to define its relationship with faith) cannot ignore the modern philosophy of science. Here, we will find, as it were, scientific reflection on the nature of science, which attempts to rise above the level of all sorts of private intuitions and experiences (for everyone of course has their own understanding of what 'science' is). We cannot even always expect leading scientists to have a completely comprehensive overview of the true nature, import and limits of scientific research. As a rule, there is no lack of individual pronouncements on the basis of scientists' own research, but to get a coherent vision we need a separate specialism. And that is how the philosophy of science has come into being, usually practised by people with both a scientific specialism and a philosophical background.[15]

One of the principal reasons for the current study is the feeling that many discussions about faith and science completely overlook the current state of affairs in the philosophy of science, and in doing so base themselves (quite unconsciously) on out-dated ideas about the nature of science. I want to try and show that it is wrong to do this, and that the relationship between science and religious faith looks very different when we bring things up-to-date.

In chapter 2, we will sketch the contours of what may be seen as the 'standard picture' of science. When we use this term, in this regard, we mean something like the common denominator of what 'we' (as the man or woman in the street) see as indicative of the nature and *modus operandi* of modern scientific endeavour. In any case, the following three ideas are central to this standard picture: (1) science develops in a progressive manner, (2) the natural sciences form the text-book example of the way in which science works, and (3) the point of departure for science is always the observation of facts which then leads to the formation of a theory. It now seems that each of these three beliefs (which are at first sight so plausible) are currently being strongly problematized and even

15 Strictly speaking it is more correct to use a plural form and refer to 'science studies', under which come new disciplines such as the history of science, the sociology of science and the economics of science. For the sake of simplicity, however, we shall use as an umbrella term 'philosophy of science'.

criticized in large parts of the modern philosophy of science. In the remainder of chapter 2, I shall try to show this in relation to (1). In order to do this, we shall provide a bird's eye view – but hopefully not without the necessary detail – of the different 'schools' in the philosophy of science that succeeded one another in the past decades and to an extent still continue to compete with each other.

In chapter 3, we will subject the second of the three elements of the standard picture of science to closer scrutiny. Is it true that the natural sciences provide the textbook example of how all 'real' sciences work or should work? It will become clear that in this case, theology is definitely not the only discipline that has a problem. Disciplines such as philosophy, law, the liberal arts, history and large parts of the human and social sciences are not in much of a better position in this regard. In this chapter we shall defend the variety that exists in scientific methods and criteria, which is itself only a function of the variety that exists in reality.

In chapter 4, we will test the merits of the third foundation of the standard picture of science. When we do this as well in the light of contemporary philosophy of science, we shall encounter the discovery of the so-called 'theory-ladenness' of human perception. Using various examples and applications, we want to track down the significance of this discovery. It will be clear that this phenomenon, though its impact should not be over-estimated, does in fact signal the end of what, for the sake of brevity, I call the TND: the Theory of the Neat Dividing-line between knowing on the one hand and believing on the other.

In chapters 5 and 6, we will then try to reap the harvest of the foregoing chapters in relation to theology. In doing so, we will make a distinction between what we shall call the older and the newer philosophies of science, considering the work of perhaps the most influential philosopher of science from the previous century, Thomas Kuhn, as the key turning point. Such a dichotomy is admittedly open to dispute, for in fact there is more of a continuum here, but we believe that its use can be enlightening. In chapter 5, we shall look at the implications of the older philosophy of science for the status of theology. In chapter 6, we shall do the same for the newer philosophy of science. Here, first of all I shall consider issues that some other commentators (especially Wolfhart Pannenberg and Nancey Murphy) have raised in this regard, and then, having learnt a great deal from them, I shall discuss my own thoughts on the matter. My intention in doing this is to show that and how theology in the classical sense of 'speaking responsibly about God' can be a fully paid-up academic discipline, and also why that is important.

In chapter 7, we turn our attention from theology to faith. This will be the point at which to look more specifically at the impact of scientific discoveries on certain specific religious beliefs. In doing so, we shall come across well-known issues from the past (Galileo, Darwin) and the present (quantum mechanics, the anthropic principle), but above all, we will again seek to provide an overview, in this case of the whole range of discussions and positions in the science-and-religion debate. In order to fulfil this task, we shall use the famous four-fold typology that Ian Barbour – one of the pioneers in the religion-and-science debate – drew up in the 1990s.[16] Although we cannot avoid making some critical comments about this classification, we consider it illuminating and helpful enough to engage with it.

At first sight, chapter 7 seems to somewhat interrupt the flow of the argument. By the end the connection will I hope become clear. The intention is to examine the extent to which our journey through the modern science and religion debate, especially as it operates in the English-speaking world, confirms or contradicts our earlier conclusions. From the fact that the majority of those who take part in this debate have moved on for a variety of reasons from the conflict model and the independence model, we shall conclude that there is in fact an important academic task for theology in this field. We can describe this task as the process of relating matters of faith in constructive and creative ways to what emerges from the modern practice of science. Because it is very important in this regard that the unique nature of faith is respected (as is the case for that matter in regard to science), we shall argue that a dialogue (or, rather, consonance) model is preferable to attempts to achieve some sort of theological or metaphysical integration between faith and science.

In chapter 8, we want to come back again to an essential aspect of theology that will have remained under-examined up to this point, namely its non-scientific side. For theology is not merely a matter of knowledge, but also one of the lived experience of faith. To use the words of the 17th century Dutch theologian, Gisbertus Voetius, it is not only a matter of science, but also of piety – and the art was for him and is for us to connect the two with each other in sensible and meaningful ways. Finally, we shall summarize our argument from yet another, somewhat different, perspective, namely the influential distinction that David Tracy made between the three 'publics' of church, academy and society for which theology has to do its work.

16 More accurately, Barbour has presented and subsequently defended and refined his proposal in different publications from 1987 to 2003; see for a discussion my essay 'Geloof en wetenschap: op zoek naar consonantie', in: Gijsbert van den Brink, Cees Dekker & René van Woudenberg (eds.), *Omhoog kijken in platland. Over geloven in de wetenschap*, Kampen 2007, 50-75.

2. Science and the idea of progress

2.1 First characteristic of the standard image: linear progress

It is partly as a result of the staggering achievements of science (some of which we briefly discussed in the last chapter) that a certain picture or image has developed in modern culture of what science encompasses and how it works. This image is, it is true to say, not particularly clear-cut, but it does have some characteristics that are particularly determinative and which influence what we think about science in an imperceptible yet significant way. In this regard, we could talk of the *standard image* of science, and in doing so give this term its own special meaning. What we mean by it here and in the rest of this book is how people, i.e. the average non-scientist (but also many practitioners of science), generally view the scientific enterprise and its results. It is of course a somewhat risky activity to sketch such an image, given that we are not going to base it on empirical-sociological material. However, even if we do not adopt such an approach, we can still talk of some global characteristics of the standard image. Of course, we would need to carry out some social-scientific research if we were to make such an image more specific, but for our purposes a rough sketch suffices. Accordingly, we shall concern ourselves in this study with three global characteristics that we come across time and again in everyday thought and discourse about what science encompasses.

The first characteristic of the standard image of science concerns the more or less constant progress ascribed to science. The idea is that of a vast range of states of affairs that are in principle all knowable. What science is concerned with is to turn what is knowable more and more into what is known. This happens, so the theory goes, in a more or less constant manner: each scientific publication that appears adds something more increase (however small it may be) to our knowledge. More and more gaps in our knowledge are filled in, until – yes, until we more or less know everything there is to know. That eschatological point lies without doubt far off into the future, and so there is still a long way to go, but we are nevertheless making steady progress towards it. It would of course be a mixed blessing when we reach this point, for then science would have to abolish itself. But this does not take away the fact that there is already speculation about a theory of everything (sometimes abbreviated to TOE – note the capital letters), in which the final secrets of the universe will be revealed. According to some people, the time when we shall reach this point is possibly not so very far off.

It was not unintentionally that we spoke a moment ago about an *eschatological* point. One does not need much imagination to see in this ideal of progress, which science is considered to fulfil, a secularised version of the Christian belief in a future that will be perfect. This 'Christian future' too finds itself at the end of a process: the process of salvation history, in which God performs his actions, and which will end up in the final divine act of judgement, of the separation of good and evil, and so in an eternal life from which all evil is banished. Now the many changes for the good that have been brought about by the tumultuous development of science have contributed to the idea that *ignorance* is an important form of evil. In the final analysis, though, it will not be God who helps to drive this evil out of the world. Rather this is something that we ourselves are and shall be involved in doing. So, whilst the content of the belief in the future has drastically changed, the structure has remained the same: we are moving towards a golden future, and as time passes we shall move ever closer to it![1] And does not science as the show-piece par excellence of this belief in progress provide the strongest proof of the truth of it?

The second characteristic of the standard image of science is a strong emphasis on the natural sciences. It is to a large extent the natural sciences, and in particular physics, that determine what may or may not be called science/scientific. The other sciences – to the extent that they can be called sciences[2] – lag behind to a certain degree, and are judged by the extent to which their methods and the measurability of their results compare with those of the natural sciences. Moreover,

1 In the first instance it was Karl Löwith (1879-1973) who described the idea of progress since the Enlightenment as a form of secularization of the content of Christian faith, in his *Meaning in History*, Chicago 1949. Löwith's thesis was heavily criticized by Hans Blumenberg in *Die Legitimität der Neuzeit*, Frankfurt a.M. 1966, who correctly concluded that if Löwith was right, then the project of modernity would be internally inconsistent. For in that case Enlightenment thought would be rooted in the very Christianity that it in fact rejected. Subsequently, Wolfhart Pannenberg has in my opinion shown why the arguments that Blumenberg used against Löwith are not convincing; see his *Theologie und Philosophie. Ihr Verhältnis im Lichte ihrer gemeinsamen Geschichte*, Göttingen 1996, 129-141.

2 Translator's note: we have to bear in mind here that the term 'science' is not exactly equivalent to the Dutch term 'wetenschap' (which is in the Dutch edition). Although the terms do often coincide, the meaning of 'wetenschap' is broader, since it includes not only the natural sciences but scholarly knowledge and practices in a more general sense. Thus, also the humanities count as sciences. In this edition, we will mainly use 'science', 'sciences', 'scientific' etc. in this broader meaning.

the extent to which something may or may not be seen as scientific is often more a question of subconscious intuition, rather than conscious reflection.

There is a kind of question and answer game in which someone says the name of a general category, and then asks the other player immediately, i.e. without giving them the slightest chance to think, to respond by giving an example of an item that may be considered to fall within that category. The questioner will already have written down what responses they expect to be given, and so can then show the person answering that they already knew what the answer would be. If for example the questioner calls out 'tools', then there is a good chance that they will immediately hear the response 'hammer'. If someone asks for an immediate example of a colour, then it seems that nine out of ten people will reply 'red'. With 'wild animal', the standard response is 'lion', with 'flower', 'rose' etc. In a similar way, we could drop in the category or idea of 'science'. It is reckoned that the most common reflex response here will be 'physics'. Of course with a little time for reflection, people will realize that biology, mathematics or astronomy are also good examples of a science. Perhaps they will even consider sociology, history or the humanities as sciences (i.e. as sound scholarly practices). But it is unlikely that any of these disciplines would be seen as a text book example of science, and so therefore that any of these would be said first in the game of reflex we have just described. The standard image of science is quite simply dominated by the natural sciences.

A third mark of the standard image is the conviction that in science we find a definitive and unequivocal methodology. This methodology can be succinctly described by the formula: from fact to theory. According to this view, things start with the scientist observing factual data. These data call for an explanation, or a theory. What scientists then do, on the basis of the observed facts, is to develop a theory with which to test the facts (the same facts, or other facts – here there is no fully-formed view). If the theory now seems to work, then it can be seen to have been proven, and another blank is filled in on the map, and the voyage of discovery in search of new facts and explanations can continue. A well-known example is of course lightning: in former times in all sorts of superstitious cultures this was believed to be an expression of the wrath of the gods, but for a long time now we have been able to explain how it is connected with electrical discharges in the atmosphere.

It will be clear to the reader that the three characteristics of the standard image that we have just described are closely related to one another. Indeed, it is above all the natural sciences which have shown the most significant progress over the past few centuries through the technology that has been developed during this period. And this progress can at first sight be seen as a linear process, whereby ever more observations have been developed into sound theories. The imaginary map on which everything that there is to know has a blank spot will be filled in more and more. Ever more observations by natural scientists are being explained, and so the domination of the natural sciences in the standard image of science continues to increase.

No doubt these three characteristics of the standard image that we have just discussed are not the end of the matter. Other characteristics might be mentioned next to them, and the standard image could be analysed in a more sophisticated way. But the idea of linear progression, the paradigmatic role of the natural sciences, and the sense of a clear, unambiguous methodology that guarantees the increase of knowledge form a remarkably strong trio. Together these three make up the core of the standard image, and are therefore responsible for the contemporary impact of the phenomenon that is science. In any case it is to a large extent on the basis of this sort of characteristic that in many situations scientists are considered to be the new priests of our culture. Once upon a time the word of the priest or the pastor (on any subject you care to mention) was considered by many as the final word in a discussion: nowadays fairly often it is scientists who get assigned that very same role. Also, whenever they speak about matters that are in no way or only slightly connected with their area of research, many people at the very least attach great importance to their words. This is of course the case once more first and foremost for natural scientists, but in their wake now sometimes hardly any less for e.g. social scientists, to the extent that, in their research and presentation, they follow the model of their colleagues in the natural sciences.

In the above, we have tried to describe the characteristics of the standard image of science in as neutral a way as possible. For it cannot be ruled out in advance that this standard image reflects reality. It could be that science does in fact manifest progressive development with almost clock-work regularity. It could be that the dominance of the natural sciences in the standard image is correct, and that as a consequence all sorts of other academic disciplines only merit the status of science in as far as they mirror the natural sciences. It could also be that the *modus operandi* of scientists is in fact rather clear and unambiguous, and that we can rightly compare it with the process of filling in more and more blanks on a map as a result of ever more facts which have been observed and explained.

In what follows, however, we shall turn our attention towards the branches of scholarly engagement that deal with analysing the phenomenon of science as carefully as possible, the so-called science studies. More particularly, we will concentrate on the philosophy of science. In doing so, it will soon become clear that recent developments in this field to a large extent problematize the standard image of science, if not totally deconstruct it. It turns out that for the most part the standard image is based not on how things work in the scientific enterprise in reality, but on the creation of some kind of image or myth around it. We want to develop this thesis and illustrate it in the subsequent chapters of this study.

In the remaining part of this chapter we will look in particular at the first characteristic mentioned above. A brief overview of the history of the philosophy of science will bring to light the fact that the widely-held view of linear progression in science is more of an ideal than a reality. In chapter 4, we will consider the narrow-mindedness that threatens to arise as soon as we ascribe to the natural sciences a regulative place in the sciences as a whole (or in our picture of science), whilst in chapter 5 the third characteristic of the standard image takes centre stage. As a result it will become clear that the idea of a simple 'from fact to theory' method does not do justice to the very complicated manner in which the development of theories in science actually takes place. This conclusion too seems unavoidable if we want to take the contemporary philosophy of science at all seriously.

2.2 From logical positivism to critical rationalism

The birth-place and time of the relatively young discipline of the philosophy of science can by common consent be pinpointed fairly accurately: the birth-place is Vienna, the time of the birth, the inter-war period. During this period, a group of scientists and leading intellectuals regularly came together in Vienna in what during the course of time became known as *der Wiener Kreis* (the Vienna Circle). Although of course reflection on science had been taking place long before, for example by Ernst Mach (1838-1916), the 'father' of this movement, it was in the Vienna circle of academics that this happened for the first time in an intensive and systematic manner.

The central figure in the Vienna Circle was the Austrian physicist and professor of philosophy, Moritz Schlick (1882-1936; in 1936, he was shot dead by a student). Around Schlick in Vienna during the 1920s, a group of logicians, physicists, mathematicians and others gathered together. Within this group (which included figures who would later become well-known such as the logician Rudolph Carnap and the sociologist Otto Neurath) there were weekly discussion evenings, that dealt with, amongst other matters, Ludwig Wittgenstein's *Tractatus Logico-Philosophicus*, which appeared in 1922. In the process, the people involved became increasingly aware that they shared a common vision of reality and of the meaning of reality. Some decided to commit this vision to paper and so in 1929 the manifesto *Wissenschaftliche Weltauffassung. Der Wiener Kreis* ('Scientific Worldview. The Vienna Circle'), written by Carnap, Neurath and H. Hahn, was produced.

This text in fact forms the Magna Carta of the first 'school' in the philosophy of science, that of so-called logical positivism. The manifesto was elaborated in a flurry of activities, congresses and publications, and had a great response internationally. In Berlin, the Berliner Gruppe (including amongst others Carl Gustav Hempel and Hans Reichenbach), which was related to the Vienna Circle, came into being, in England Alfred Ayer was a great champion of the ideas of the Vienna Circle, in Scandinavia Arne Naess and others got involved, whilst in Münster Heinrich Scholz, who later became known from his debate with Karl Barth about the scientific character of theology, showed sympathy towards it. This process of internationalisation increased in the second half of the 1930's when many members of the Vienna Circle (especially those who had a Jewish background), under pressure from the rise of Nazism, were forced into a sort of diaspora, above all to the United States. Although this meant a rather abrupt end to the Vienna Circle (its heyday was in the period 1930-1935), the ideas which had been developed in it gained that much more influence on both sides of the Atlantic. All in all, logical positivism was to remain the dominant movement in the philosophy of science until well into the 1960's. And beyond the philosophy of science, its influence is still clearly tangible: the standard image of science that we discussed in the previous section can to a large extent be traced back to logical positivism.[3]

After this short overview of the fascinating history of the Vienna Circle and logical positivism, if we now turn to look at the intellectual content of it, then we see that the title of the manifesto published in 1929 which we referred to above seems particularly well chosen: the logical positivists are concerned with a *scientific worldview*. In the manifesto this type of worldview is contrasted with an outlook on life and vision of society that are to a large extent characterized by theological and speculative-metaphysical ideas. It was felt that these sorts of ideas were unjustifiably gaining ground as some people in German-speaking lands after the First World War had adopted a kind of policy of restoration, a 'back to earlier times' movement. It was this type of idea, which was seen as authoritarian and hopelessly conservative, with an emphasis on traditional values such as authority, obedience, work, piety etc., that the Vienna Circle wanted to oppose. For this reason, its members tried to develop the alternative of a completely *scientific* worldview.

Fundamental to this is the conviction that only science can determine what is true and what is not. In order to fulfill this task, however, science has to be completely defined by empirical knowledge, i.e. by knowledge stemming from our sense experience of reality. This is why we speak not only of logical positivism, but also of logical *empiricism* (although these two terms are not strictly speaking

3 The standard image we have discussed is in fact a sort of diluted version of the so-called 'received view', a reconstruction of the logical-positivist picture of science. For a presentation and critique of this 'received view', see F. Suppe (ed.), *The Structure of Scientific Theories*, 2nd edition, Urbana 1977, 1-232.

completely synonymous: an empiricist will agree with the epistemology of positivism, but not necessarily with its ideological claims).[4] Taking basic sense(-perceptible) data as our point of departure, we can build the house of science, as it were, as a one-piece building. And the task of philosophy is to make the unity of this house visible by means of logical analysis, and by reducing all truly scientific statements to their empirical substratum. All possible knowledge that is acquired in the various scientific disciplines can thus be presented as inter-connected, based on the hard bedrock of indisputable knowledge: the firm foundation of the data of direct perception, the *sense data.*[5] Surely we need a special language for this, as our everyday language is not precise enough to account for it. Therefore, our everyday language has to be transformed with the help of symbolic logic into an abstract scientific language.

And here we have what is attractive about logical positivism: an all-encompassing empirically-rooted 'science of unity' (*Einheitswissenschaft*) with physics as its model, formulated as strongly as possible in an abstract scientific language, as a powerful antidote to all the unprovable claims of theology and metaphysical philosophy. By formulating science in this way, it can, so it was hoped, become the successor to theology as the provider of a complete view of life. For all that we do – parenting, education, politics, economics etc. – can ideally be formulated scientifically from this starting point. We can clearly see how great the expectations were from the following quotation:

> Neatness and clarity are striven for, and dark distances and unfathomable depths rejected. In science there are no 'depths'; there is surface everywhere: all experience forms a complex network, which cannot always be surveyed and can often be grasped only in parts. Everything is accessible to man; man is measure of all things ... The scientific world-conception knows no *unsolvable riddle*.[6]

4 Hans Reichenbach considered himself to be a logical empiricist but not a logical positivist for another reason: that he rejected the verification principle (which we shall discuss below); see Martin Curd and J.A. Cover, *Philosophy of Science. The Central Issues*, New York 1998, 427, 1285. Both terms are however often used as synonyms.

5 Carnap later described the epistemological convictions that he had in 1929 as follows: 'We assumed that there was a certain rock bottom of knowledge ... which was indubitable. Every other kind of knowledge was supposed to be firmly supported by this basis and therefore likewise decidable with certainty.' Cf. Paul A. Schilpp, *The Philosophy of Rudolf Carnap*, La Salle (Ill.) 1963, 57. Compare also Carnap's interesting account of a conversation on this matter with Einstein (*ibid.*, 38).

6 From the English translation of *Wissenschaftliche Weltauffassung*, quoted in Otto Neurath, *Empiricism and Sociology*, Dordrecht 1973, 306 (italics in this text).

If we look closely, we see all sorts of elements from older philosophical movements return in the logical-positivistic programme. The idea of the empirical world as a basis of all our knowledge, for example, has its roots in the classical empiricism of David Hume, and was brought into the philosophy of science by the man who inspired the Vienna Circle, the aforementioned, Ernst Mach, who had taught Moritz Schlick. And the idea of science as the successor to theology had earlier been proposed by Auguste Comte, who saw that, after a theological and metaphysical stage, in his lifetime a scientific or *positive* phase of history was breaking in, which would bring nothing but true progress. The ideas of Comte were generally referred to with the term positivism, and the logical positivists (or *neo-positivists*, although this term is sometimes also used in a broader sense, i.e. including the Popperians whose ideas we will discuss below) differentiated themselves from him in that they linked the idea of starting from 'positive' scientific facts with a strongly logical approach, which was different from what Comte himself did.[7]

This logical approach becomes clear from an important element in logical positivism that has a direct connection with its worldview, and that particularly interests us now, viz. the so called verification principle. This principle asserts that if we want a statement to be called scientific with justification, it must be verifiable. And a statement is only verifiable when it can be directly traced back to certain sense data, i.e. to sense-perceptible facts. It was believed that only in this way could language be 'cleansed of all metaphysical contamination'. Theological or speculative-philosophical castles in the air cannot of course be reduced to statements that can be proven with sensory experiences or actual observations. So, according to the verification principle, they are immediately found wanting, and are to be considered unscientific. For it must be possible for a statement (or phrase, or expression, or system) to be derived in a logically sound way from empirical facts if it is to be given the predicate 'scientific' with any justification.

But this is not all. We could imagine that theologians and philosophers would not find it so annoying to see their ideas characterized as unscientific. There is in fact more to life than science and science does not in and of itself have a monopoly on truth and wisdom. The logical positivists, however, added a further condition to the verification principle: a statement that cannot successfully be tested using the verification principle is not only unscientific, but it is also without meaning. Taken in its strictest sense, this means that we cannot even say that such a statement is untrue, for whether a statement is true or not can only be

7 For summary descriptions of the ideas of Hume and Comte, see, G. van den Brink, *Oriëntatie in de filosofie*, 203-208, 277-282; for the relationship between positivism and logical positivism, see esp. 316.

determined when we know what the particularly statement means. In that sense, the verification principle is a criterion of meaning rather than a demarcation criterion. Since it is impossible to say what a statement that is not empirically verifiable means, it is meaningless to make such statements. And whoever does make such statements in fact manages to say nothing, and would do better to keep quiet. Only by making this condition can we prevent all our language from being contaminated by religion and metaphysics. And it is from this that we get the robust formulation of the verification principle such as that given by Alfred Ayer,

> We say that a sentence is factually significant to any given person, if, and only if, he knows how to verify the proposition which it purports to express – that is, if he knows what observations would lead him, under certain conditions, to accept the proposition as being true, or reject it as being false.[8]

This logical-positivistic verification principle can be seen as a heroic attempt finally to separate the wheat from the chaff. Once and for all it should be clear that we can not just say anything willy-nilly, but have at all times to be prepared to lay our cards on the table. It is all the more dramatic, however, that this attempt to divide science and humbug, meaning and nonsense from each other must by common consent be seen as a *failure*. The verification principle turned out to be unable to fulfil the role intended for it. It quickly became clear that far too many problems were associated with it. In particular, the severity of the principle came back like a boomerang on logical positivism itself: all sorts of statements that were considered as full of meaning and true, turned out to be emperically unverifiable.

In some cases, a way round was found, although opinion soon differed on how this could be achieved. A problem that soon arose concerned for example the status of mathematical and logical expressions (such as 2+2=4). These can in fact only be reduced to agreements established once and for all on the meaning of the symbols concerned, and cannot be reduced to empirical observations. For this reason, some scholars suggested that it simply had to be admitted that logical and mathematical expressions are in fact meaningless. Others instead believed (following on naturally from what Hume and Kant had already previously proposed) that it is better to create a sort of special status for this type of statement: we are dealing here with *analytical* statements, the truth of which is already established thanks to an agreement on the meaning of the symbols concerned. In this case, empirical proofs are therefore not necessary, but we can take the correctness of these statements for granted without such proofs. People had of course to believe that there was such a special status, but in any case this could be reasonably argued for. However, one had to allow for exceptions to the rule

8 A. J. Ayer, *Language, Truth and Logic*, Harmondsworth 1971 (original edition 1935), 16.

anyhow, for example for the verification principle itself. This itself is in fact not empirically verified either, and it is difficult to see how it could be in any case.

Other problems turned out to be even more serious. It is going too far to try and sum these up here comprehensively,[9] and it will suffice to limit ourselves to one of the most persistent and also most illustrative of these, the so-called *problem of induction.* This concerns the status of so-called inductive generalizations. By this we mean general statements that come about by combining all sorts of separate observations which are mutually supportive. Well-known examples of such inductive generalizations are the empirical laws of nature. Generally, it seems, the most basic form these can take is a statement such as 'all entities A behave in manner B under conditions C'. For example, all metals expand on heating. The textbook example of an inductive generalization is even more simple: all ravens are black.

Such an extremely simple assertion does not satisfy the verification principle, and for this reason is said to have no meaning. Even if a black raven has been observed many thousands of times and furthermore no raven has ever been seen which is not black, then it still will not have been proven empirically that all ravens are black. In order to be able to verify the statement all ravens will have to be seen – and that is in fact impossible. Even if someone had managed to see all the ravens in the world, then it would still be impossible to know this and thus to rule out the possibility that there were in fact still some ravens that had not been seen. However many ravens someone saw, they could never move from the proposition 'the x ravens that I have seen are black' to the conclusion that 'all ravens are black'. And the latter proposition cannot be empirically verified in any other way.

If we reflect on the fact that a similar problem also attaches to all the laws of nature,[10] then we see that this is not simply a matter of some sort of over-sophistication of secondary importance. On the contrary, the most celebrated and influential, most fundamental, most comprehensively convincing and conclusive of all scientific statements, i.e. the laws of nature, can, according to logical positivism, not strictly speaking be thought of as scientific! And given that there are only two ways to go for our assertions, they should then be considered as belonging to the detested category of metaphysical speculation. But such a conclusion of course sounds completely counter-intuitive, and not only to logical-positivist ears.

9 For brief overviews of the problems of logical positivism in general and of the verification principle in particular, see the entry 'logical positivism' in a contemporary dictionary or encyclopaedia devoted to philosophy, e.g. Robert Audi (ed.), *The Cambridge Dictionary of Philosophy*, Cambridge 1995, 445f.

10 This applies not only for the straightforwardly empirical but also for the so-called theoretical, such as the laws of Newton's classical mechanics. Here too we are dealing with universal statements (in other words with statements of the type '*all* x...').

In order to avoid these and other, similar awkward conclusions,[11] during the 1930s the verification principle was somewhat watered down by reducing it to a so-called *confirmation* principle. This asserts that we are entitled to consider that a statement is universal, whenever it has been confirmed inductively (i.e. in all sorts of separate situations) often enough. If someone has enough examples of black ravens, and has never seen a white or green raven, then at a certain moment they can take the plunge and consider that they are justified in making the statement all ravens are black. Such a statement does not then need to be seen as meaningless. More generally, a universal statement can be considered to belong to the language of science, when it has been sufficiently validated or *confirmed* by empirical evidence. It is admittedly true that strictly speaking it will not have been shown that the relevant statement is true. But there are nevertheless good grounds to believe that it is highly *probable*.

In this way, the original strict verificationism was replaced by the more moderate confirmationism, also known as probabilism.[12] At first sight the replacement of the verification principle by the confirmation principle seems a completely reasonable concession to make, given the rather inordinately strict criteria that the original logical positivism placed on statements. Sometimes you have to make a little sacrifice for a greater good. But we should not underestimate the crucial significance of this sacrifice. However hard it may be to swallow, in a confirmationist context we can not avoid letting unproven *conjectures* creep into the foundations of the building that is science. It is admittedly true that we are dealing here with conjectures that as a result of a good deal of empirical testing may be considered to have a high level of probability. But even so, from now on science is no longer built on the foundations of unquestionably true statements.

In other words: so-called foundationalism, which is a common thread that runs through the history of Western thought, and of which logical positivism is the most sublime expression, is beginning to show cracks that in the long run we shall not be able to cover up.[13] Nowadays in epistemology, this foundationalism is generally considered, certainly in this powerful and pure form, to have become

11 Not only universal, but also individual statements of observation seemed (for other reasons) not to be strictly speaking completely verifiable.

12 For the shift from the verification to the confirmation principle, see Carl G. Hempel, 'Empiricist Criteria of Cognitive Significance', in his *Aspects of Scientific Explanation*, New York 1965, 101-122 (and compare 3-46).

13 Much has been written about foundationalism in the last decades, especially in American philosophy, and most of the literature has been critical of it. See entries for this term in recent philosophical dictionaries, and the seminal collection of essays, Alvin Plantinga and Nicholas Wolterstorff (eds.), *Faith and Rationality*, Notre Dame 1983, in particular the contributions by both of the editors.

obsolete. Philosophers discuss what the rightful successor to it should be (coherentism, relativism, perhaps a less rigid form of foundationalism, or whatever), but they no longer discuss the question of whether there should be a successor or not. Now that the passionate search for a complete set of absolute certainties that can be established in a rational way has failed, also in logical positivism, it is agreed that classical foundationalism is unsustainable.[14]

With the shift from verificationism to confirmationism, the watertight distinction between provable (i.e. verifiable) scientific and unprovable non-scientific statements is at risk. For this reason, on closer inspection the move to confirmationism can not be seen as a little detour in the otherwise constant progress of the logical-positivistic programme. In fact, it went to the very heart of that programme. It was then an enormous setback for the founding fathers of logical positivism to have to accept this shift and some of them would without doubt not have been ready to do this.[15]

However, it would even get worse. For far from offering a safe haven, the confirmation principle turned out to bring with it its own set of problems. Obviously, the *first* problem was that it was based on the uniformity of nature. Its basic idea is that when a certain pattern is repeated and empirically confirmed often enough in nature, we can accept that this pattern also occurs in all those places where it has not been observed. And consequently it is believed that whenever we do not observe it, nature in fact behaves as it does when we do observe it. Indeed this is a basic premise within logical positivism (and also as it happens within the modern scientific enterprise). But on what basis can we justify such a premise? In any case, it has not been verified. It is not even possible to verify it, for as soon as we try to verify it, we will no longer be dealing with nature as unobserved.

So we have to retreat to the position that it is at least very *probable* that nature behaves in a uniform manner, for example because this very often turns out to be the case. The view that nature is uniform has, to put it another way, a high degree of confirmation. But this assertion does not show beyond doubt that the confirmation principle is right. Rather it is an attempt to stop one leak (the confirmation principle which does not lead to absolutely certain knowledge) with another (the uniformity of nature which is just as difficult to verify). For clearly, we are using an inductive argument (nature constantly seems uniform) to support the

14 We have seen this search time and again in the history of Western philosophy from the beginning of modernity, e.g. with Descartes, Husserl, the early Wittgenstein and so also in logical positivism.

15 So, Nicholas Wolterstorff, *Reason within the Bounds of Religion*, Grand Rapids 1976, 34. (There is a second expanded version of this book published in 1984, but I prefer the original edition).

position that we are entitled to adduce inductive arguments.[16] The only way to solve this riddle is simply to stipulate that the confirmation principle, just as we saw in relation to the verification principle, is in fact an unprovable, yet valid, point of departure. To justify this position, we could for example point to the fact that it works, that in other words we really make progress in our attempts to separate the scientific sheep from the metaphysical goats by taking the confirmation principle as our starting point.

At this juncture, though, a *second* problem arises: the confirmation principle turned out to be particularly difficult to apply. For how does one decide when a certain statement or theory has reached a *sufficient* degree of empirical confirmation to be rightly considered as scientific? Rudolph Carnap, mentioned above, one of the most influential members of the Vienna Circle, devoted all his life to trying to provide a definitive answer to this question. He did so by looking for a way in which the degree of confirmation could be quantified. This would have as a consequence that it could for example be asserted: universal statement A is confirmed by proof B to the extent that A has a probability (that is a degree of confirmation) of 0.8 – i.e. that the chance that A is true is 80%. But in spite of all their efforts, Carnap and others did not succeed in quantifying the confirmation principle in this way. As a result, the principle could not be put into practice. One had to make do with the general conclusion that as a certain statement or theory is confirmed more and more often by empirical facts (for example as all sorts of predictions that are made from it turn out to be true), the chance that this statement or theory is correct becomes greater. Thus, confirmationism implies that in principle we will always have to continue the search for further corroboration of our statements and theories. If it seems that such empirical confirmation is totally absent, then we have to conclude that the relevant statement or theory is cognitively meaningless.

Perhaps we could have lived with this rudimentary version of the confirmation principle, were it not for the fact that just at this point a *third* problem arose. This third problem seemed so insidious and stubborn that it would turn out to be fatal for confirmationism (and in fact for the logical positivistic programme as a whole). The honour for having brought this to light, with, as it happens, unceasing zeal, goes to one man: the British philosopher, born (yet again) in Austria, *Karl R. Popper* (1902-1994).[17] Popper had studied philosophy and psychology in Vienna

16 On this point, compare Wolterstorff, *Reason within the Bounds of Religion*, 35f.

17 Popper saw himself, as is evident from his autobiography (*Unended Quest*, London 1976, 69), as the person who dealt the fatal blow to logical positivism, as a result of his critique of the verification principle. We should also note here, though, that this verification principle in its strong form had already been abandoned by this time within logical positivism. This does not take away the fact that Popper did deal the fatal blow indeed.

in the period in which the Vienna Circle came into being. He thus knew, as it were, from the off what was what. Popper had an ambivalent relationship with the members of the Circle. On the one hand there was mutual respect, but on the other over time Popper became increasingly critical of the logical empiricism of the Vienna Circle.

In fact Popper's criticism concerned first and foremost the epistemological side of the movement. The political ideal of social reform based on a uniform scientific method was shared by him, but Popper was less critical of classical metaphysics in this connection. The great philosophers from the past were definitely concerned, according to him, with real problems, although they were not scientific problems as such. Popper was convinced of the structural *incompleteness* of science, and pointed to the fact that all sorts of scientific theories had had preliminary stages which were metaphysical in nature (for example the idea that light consists of particles). These metaphysical ideas cannot simply be rejected as meaningless. Surely Popper believed that metaphysical theories must in principle also be testable, and that they therefore ought to be related to scientific research. But when it came to his own metaphysical views (for example that the universe is 'open' in the sense of being indeterministic) Popper did not follow this rule: he simply could not think of any single scientific development that could disprove this view.

In spite of the fact that Popper therefore had a higher regard for non-scientific forms of reflection, just like the logical positivists he looked for a way of making the boundary between scientific and non-scientific knowledge claims more clear-cut. In this regard, Popper spoke of the problem of *demarcation*. It gradually became evident to him that the confirmationist criterion was inadequate for solving this problem.

On a biographical note, Popper came to this conclusion as a result of personal life-experiences. These stemmed from the time that he worked in Vienna as a volunteer for the famous psychologist, Alfred Adler, in his clinic for children with developmental problems. During this time, it occurred to him that Adler always explained the problems of these children in one and the same way: that they arose as a result of inferiority complexes. Whatever behaviour Popper presented to Adler and his assistants, every time Adler managed to make a link with his theory. And if certain observations did not seem to agree with this theory, then he always knew how to devise an *ad hoc* argument which brought these observations back into line. And so it seemed as if his theory was continually being verified and that it was therefore becoming stronger and stronger. Whenever Popper asked Adler how he could be so sure that a particular behavioural problem had to be explained by his theory regarding inferiority complexes, he referred to his experience of thousands of cases. But it was just this fact that Adler (and Popper had the same impression about Freud with his psychoanalytic theory) seemed to be able to explain every single behaviour as confirmation of his theory that made Popper suspicious. Was the apparent strength of the theory – so many positive examples! – not in fact its great weakness, because no event could be thought of which could *not* be explained as a confirmation of his theory?

Popper had a similar experience with Marxism, which was very popular in Vienna for a time, and to which Popper subscribed for a short period. Popper observed that Marxists let nothing stand in their way in order to achieve their goals and he concluded from this that their theory concerning the necessary anti-capitalistic turn that history would take had to be particularly strong.[18] However, with a shock he realized one day that Marxists managed without difficulty to explain every event in the newspapers as a confirmation of this theory. If they did not manage to see it as a step towards the fall of the capitalist system with one approach, then they would try another. So, every event brought the outcome of the relevant prediction a step closer from a Marxist perspective. And this theory too seemed to be confirmed by a whole host of facts. But at a certain moment behind this apparent strength Popper also very clearly saw the unmistakable weakness. For is it in fact possible to *critique* the theory on the basis of certain events? And if it is not, if instead everything that happens can easily be reconciled with the theory, then does the theory say anything at all about the reality in which we live, or is it instead an illusion?[19]

In contrast to this, Popper reminds himself that at the beginning of the 1920s, he read somewhere a popular essay on Albert Einstein's theory of relativity. How different Einstein's way of working was compared to that of Adler, Freud and the Marxists! Einstein's theory of relativity had a far higher level of abstraction; but Einstein had dared to attach a concrete, testable prediction to it concerning a previously unforeseen effect that would have to occur if his theory was to be valid. This stated that rays of light that come close to a heavy mass such as the sun do not move in a straight line, but are to an extent bent by this mass. It is true that the extent to which they are bent is not great, but it is still clearly visible, and certainly more than Newton's theory of gravity allows for. Einstein now stated in no uncertain terms that if this phenomenon did not conform to his prediction, then his general theory of relativity should be rejected.

18 It is interesting that Marxist theory, in contrast to that of Adler, dared to make certain *predictions*. It was precisely for this reason that Popper ascribed scientific status to it for a while.

19 See for example, Karl Popper, *Realism and the Aim of Science*, London 1983, 162-174 and compare the longer discussion in his *The Poverty of Historicism*, London 2002 (original edition London 1957; by 'historicism' Popper was alluding above all to Marxism).

As it happens, no one had ever noticed such a deviation before. This was in particular the result of one complicating factor, i.e. that it could only be confirmed when there was a total eclipse of the sun. Without such an eclipse of the sun, rays of light from other celestial bodies that are close to the sun merge with the sunlight so that the various rays can no longer be distinguished. For this reason at the suggestion of the British physicist Alfred Eddington in 1919, two expeditions went to places in West Africa and Brazil so that in May of that year they could witness a total eclipse of the sun. But because of the mediocre quality of photographic techniques (and due to bad luck: on the crucial day the sky in West Africa was overcast) the expeditions did not deliver the decisive triumph for Einstein that was attributed to him in the press. Not until 1922, when a new expedition went to the Pacific coast of Australia, did the now much more accurate measurements agree once and for all with Einstein's prediction. As a result Einstein's general theory of relativity became that much stronger, and the scientific community was now totally convinced of its validity.

According to Popper, who always continued to see Einstein as his great hero, it was in the very *fragility* of Einstein's theory that its strength lay. For whereas other scientists and pseudo-scientists sought to do nothing other than continually confirm their theories, Einstein did precisely the opposite: he formulated as closely as possible which empirical conditions would *disprove* his theory. And so he tested his own theories critically, instead of uncritically trying to bring all sorts of observations into agreement with them. And it is just this critical attitude towards his or her own theories (and those of others) which characterizes the true scientist. Such a scientist, as Popper argued in his *Logik der Forschung*,[20] rightly seeks out risky experiments, 'crucial tests' that can totally undermine the theory, instead of observations that can only confirm it.

In other words: it is not in the possibility of confirmation, but rather in the possibility of rejection or *falsification* that the criterion for distinguishing between true science and pseudo-science (Freud, Adler), ideology (Marx) and metaphysics can be found.[21] Only by consciously exposing our theories to possible critique and testing them in a process of trial and error[22] can our scientific knowledge of reality

20 Karl R. Popper, *Logik der Forschung. Zur Erkenntnistheorie der modernen Naturwissenschaft*, Vienna 1935.

21 Popper's 'falsificationism' only came to a wider audience through the English translation of his *Logik der Forschung*: Karl R. Popper, *The Logic of Scientific Discovery*, London 1959. The Groningen philosopher and psychologist Michel ter Hark argues in his Ph.D. dissertation, *Popper, Otto Selz and the Rise of Evolutionary Epistemology*, Cambridge 2004, that Popper appropriated the core of his falsification thesis from the German-Jewish psychologist Otto Selz (1881-1943).

22 Compare the title of Popper's second extensive book about the philosophy of science: *Conjectures and Refutations*, London 1963.

increase. This is precisely what happened by means of the 1922 experiment described above: it has made us a little more knowledgeable. For this reason scientists must strive to formulate their theories and the predictions that flow from them as specifically and precisely as possible. A meteorological prediction such as 'It will rain in Europe this week' is much less scientific than the prediction 'It will rain tomorrow lunchtime in village x'. Of course it always rains somewhere in Europe at least once every week, so the first 'hypothesis' can hardly be falsified. But with the second by contrast it is completely clear what we must do in order to test whether or not it is correct: make sure that we are in village x tomorrow lunchtime.

When scientists work in this way, it does not however mean that they can discover more and more *truth*, that they can definitively fill in more and more blanks on the map of our (potential) knowledge. For clearly it is always possible that a theory emerges unscathed from one or two experiments that could possibly have undermined it, but then fails in a third. For this reason Popper prefers not to speak of 'confirmation' when a theory has survived an attempt at falsification, even if the theory has in fact been strengthened or corroborated by it. For it can still turn out to be incorrect. It is even impossible to tell how many tests a theory must emerge from successfully before we are allowed to call it true. In this connection Popper adopted a saying of the German mathematician Hermann Weyl: „Nature calls out a loud 'no' (i.e. whenever she falsifies a theory), but her 'yes' is almost inaudible."

Therefore, although a theory that is subject to experiments that are aimed at falsification can surely be called scientific, it is not yet by definition also *true*. For we can always imagine that a theory that has successfully come through a good number of experiments may, in the course of time, be confronted with a test which it does not come through successfully. As a result, at the end of the day truth (in the sense of fully justified belief) is unobtainable in science. This does not mean, however, that we should not strive after it! Popper saw truth as a regulative ideal in the Kantian sense: a perspective that provokes us to undertake research and which therefore must be pre-supposed as a given, but about which we cannot say anything else. Later in his career Popper tended towards the idea that we gradually move ever closer to the truth.[23] Strictly spoken, however, the only thing that we can know for sure is that falsified theories are *not* true. For this reason Popper spoke of the *growth* of our scientific knowledge rather than of progress, since the latter term suggests that we more and more come to know what is the case, whereas in fact we only know more and more what is not the case. For as we saw, it is very difficult to say to what extent theories that have not been falsified are true. From a practical point of view, however, at a given

23 For Popper's idea of 'verisimilitude' see his *Conjectures*, 219, 233-237

moment we are certain enough of our beliefs to dare to make such a claim. It is like driving piles into a marsh: we can always discuss whether we should hit them one more time into the ground. But solid rock – the rock which Carnap initially sought with such passion – will never be reached. At a certain moment instead it makes sense to stop driving the piles further into the ground and to begin building.[24]

All in all Popper thus showed himself to be a 'fallibilist', i.e. someone who is very aware of the structural fallibility of human reason. In this respect, his experiences with Adler and the Marxists left an indelible mark on his thought. Many of our ideas do not hold water, even if they sometimes seem so rational. Even mature scientific theories (such as for example those we have from Newtonian physics) go unchallenged for centuries, until they finally they run up against their limits and are replaced by stronger alternatives (such as Einstein's theory of relativity). This growth in our scientific knowledge only occurs when we dare to place our theories and hypotheses under the strong critique of empirical reality. In other words, Popper was a *critical* rationalist – and so his position in the philosophy of science should enter the history books under the heading 'critical rationalism'.

Popper also showed himself to be an independent and original thinker in other areas apart from the philosophy of science; in particular he had outspoken ideas about the way in which our society could best be organized.[25] Just as in his epistemology, so too in his social and political philosophy openness to criticism and the rejection of closed systems which only validate themselves play an important role. But it is as a philosopher of science that Popper became most influential. He is generally seen as one of the most important philosophers of science of the twentieth century. Scientists in many fields and other intellectuals accept his falsificationism as the legitimate successor to logical-positivistic confirmationism. And right up to the present day many are happy to turn to Popper in order to give an account of what science really is and how it proceeds. His vision of this has been very inspirational and to a certain extent still is.

At the same time, we must say that Popper's contributions to the philosophy of science have become the object of incisive criticism which cannot be ignored. Whilst those working in the same field as Popper on the one hand stood on his shoulders, on the other hand they meticulously pointed out the weak points in his views. And once again it was not just questions of secondary importance, but the core of Popper's programme, i.e. the falsificationist criterion of science which was at issue. This took place, however, within a much broader framework: that of the development towards a whole new phase in the philosophy of science. But

24 Popper, *The Logic of Scientific Discovery* (ed. New York 1965), 111.
25 Compare K. R. Popper, *The Open Society and Its Enemies*, 2 volumes, London 1945.

before we consider this newer philosophy of science (in section 2.4), we shall spend a little time looking at some interesting precursors to it (section 2.3).

2.3 The thesis of Duhem and Quine

Let us look again at Popper's favourite example of how science should work: Albert Einstein deducing a very specific, testable hypothesis from his extremely complex and abstract theory of relativity. But let us assume for a moment that the experiment done by Eddington that we discussed above had turned out differently. In other words, that rays of light that come close to the sun do not (under the influence of the mass of the sun) bend with the curvature that Einstein predicted. If this happened, as Einstein himself admitted, his theory would have to be rejected. Here, though, a difficulty arises which the French Roman Catholic physicist and philosopher Pierre Duhem (1861-1916) had already pointed to early in the twentieth century and which was later worked on and defined further by the Harvard logician W. V. O. Quine (1908-2000). For, exactly which *part* of Einstein's theory of relativity would have to be rejected? This question which seems so simple turned out to be extremely difficult to answer.

Duhem had already noticed in 1906, that even a relatively simple scientific theory such as Boyle's Law (the pressure of a gas, at a constant temperature, is inversely proportional to its volume: i.e. when the pressure increases, so the volume decreases etc.) in reality consisted of a complex combination of conscious and unconscious assumptions. Some of these are essential, others more of secondary importance, but all play a role. So, for example, when Boyle's Law is being tested, there is an assumption that all the instruments are in working order, that the measurement techniques are correct, that the mathematical methods being used are valid, etc. As a result, whenever there is a falsifying experiment, the question arises where exactly the fault lies. The hypothesis which is being tested may be incorrect, but it is just as possible that something is wrong with the other assumptions. In other words: falsification tells us that something is wrong, but not exactly what is wrong. For this reason, it is difficult to speak in this regard of a 'decisive experiment'.[26]

However, one might object that this problem is not in fact intractable. In order to solve it, we have only to reduce the theory to all its separate assumptions and hypotheses, and then test each of these one by one. Duhem though stated (and in

26 Compare P. Duhem, *La théorie physique, son objet et sa structure*, Paris 1981 (1906), translated into English as Pierre M.M. Duhem, *The Aim and Structure of Physical Theory*, Princeton 1991 (1954). (The later editions indicate the enduring importance of this work).

doing so he differed from the view that Popper would later adopt) that in practice this is impossible. Even if we were aware of all the separate assumptions, or could become aware of them – and Duhem denied this was possible – then we still could not entirely isolate them from one another. For clearly, a hypothesis that we want to test is deduced from a certain theory, and this theory consists of a set of closely inter-related statements. In addition, there are all sorts of assumptions about which testing procedures are correct. Now we can of course agree about which of these statements and assumptions (for example regarding which calculating methods to use) we believe to be true. And indeed in scientific practice there are many of these types of more or less generally accepted conventions in operation. But obviously the truth of a certain theory cannot be shown by appealing to conventions that we ourselves have brought into being ....

Whereas Duhem limited this alarming analysis to the realm of physics, when in 1951 Quine referred back to Duhem's ideas, he did not limit himself in the same way. Quine in fact argued in an article which became famous,[27] that not only can a hypothesis not be tested in isolation but also its contents cannot be articulated in isolation. This has to do with a feature of all our statements about the world in general (and so not only statements pertaining to the natural sciences): by definition they take their meaning from a greater whole. Every empirical statement consists of certain concepts, and in order to grasp the meaning of these concepts, we need yet more statements. These other statements rely in turn on certain other assumptions and interpretations etc. So, all in all, we can speak of a complete network of assertions which mutually influence and strengthen each other, a network[28] which according to Quine only skirts the edges of our experience of reality. This phenomenon is sometimes referred to as 'semantic holism': the meaning of a statement can only be understood in the context of a greater whole. And the meaning of a theory can therefore not be reduced to being the sum of the meanings (or actual contents) of its constituent parts. Rather, the meaning of a theory is, as it were, 'spread out' or diffused over the whole and closely connected with ideas and assumptions *outside* the theory.

And so we have what is called 'the thesis of Duhem and Quine': there can be no experiment that decisively disproves a theory, because it can never be determined exactly what part of the theory (or of the auxiliary hypotheses which must be made use of in the experiment) is being disproved. Repair work in order to avoid the negative result of the experiment can usually be carried out in different

27 W. V. O. Quine, 'Two Dogmas of Empiricism', most accessible in: *From a Logical Point of View,* Cambridge (Mass.) 1953, 20-46.

28 Quine further developed this metaphor of a network of knowledge in the book that he wrote with J. S. Ullian, *The Web of Belief*, New York 1979.

places in the network of the theory – and in principle it will be impossible to determine objectively which repair is the correct one. According to Quine not even the most generally accepted statements are a priorily immune from possible revision. At first sight, the thesis of Duhem and Quine hardly seems plausible. For had not science since the sixteenth century onwards known how to unlock numerous mysteries in a skilful and systematic way, and how to solve numerous problems one after another? But as things have turned out the Duhem-Quine thesis has continued to influence debates in the philosophy of science. To be sure, it did not immediately have far-reaching consequences for everyday matters on the scientific shop-floor, but that does not mean that it did leave things undisturbed.

A clear example that may serve as an illustration here is the extensive discussion at the end of the nineteenth and beginning of the twentieth century about the way in which light moves through the ether in our solar system. The 1887 Michelson-Morley experiment, which became very well-known, had shown that the ether which surrounds the earth does not move with respect to the earth. This result was problematic, for the existing theory (posited by H. A. Lorentz) that was tested with the help of this experiment, was based on the idea that this ether does in fact move with respect to the surface of the earth. In addition, this theory consisted of a whole number of other assumptions, such as the one that ether has a certain speed, that light interferes with it in certain ways etc. An extensive literature consequently arose in which the most ingenious solutions were put forward to explain the problematic result of the Michelson-Morley experiment. It was not easy to judge these solutions on their merits: some of them could be tested, but it was possible to get round some of the less successful results. Indeed several theories did accord with the facts. It was only in hindsight that it became clear where the real 'problem' was to be found, viz. in an implicit assumption that was so fundamental to the whole theory that for a long time nobody had wanted to cast doubt on it: the assumption that there really *was* something like ether![29]

The interesting question that then of course arises is whether the way in which this conclusion was reached was a rational one, or rather a matter of pure coincidence. Or to formulate it more precisely: the question is whether this conclusion can be rationally justified in retrospect. For clearly, it is a generally recognized fact that very many discoveries in science are made purely by chance (there is even a special term for this: *serendipity*, i.e. the phenomenon that during scientific

29 We borrow this example from A. E. McGrath, *The Foundations of Dialogue in Science and Religion*, Oxford 1998, 91-92. Regarding the thesis of Duhem and Quine also see McGrath's more introductory book, *Science and Religion. An Introduction*, Oxford 1999.

research important discoveries are made completely by coincidence whilst something else is being looked for).

The logical positivists already knew about such coincidental discoveries. However, they made a strong distinction between what they called the 'context of discovery' and the 'context of justification' of scientific research.[30] In the former, all sorts of contingent circumstances and factors of a historical, psychological, social or economic nature come into play. If an important scientist was sick on a certain day, or perhaps just less diligent, then the discovery would not have been made and the development of his or her discipline would perhaps turn out completely differently. However, according to the logical positivists such contingent factors are irrelevant from the perspective of the philosophy of science. What matters is the *logical justification* of theories which have been developed in whatever way, a justification that must make it clear that these theories are in fact correct. That is what must be demonstrated. Popper appropriated this distinction from the logical positivists, but the question is whether it can be maintained given the Duhem-Quine thesis. Is there really an 'objective' way of showing that the abandonment of the whole ether-theory was the only correct response to the complete set of all relevant observations? Is it possible to logically justify this move as being the only rational one?

We shall not go into this any further at the moment, but we shall see that just this sort of question returns with an unparalleled sharpness in the following phase of the development of the philosophy of science. During this next phase, the clear distinction between 'context of discovery' and 'context of justification' becomes more and more challenged. The thesis of Duhem and Quine can be seen as an important precursor to this phase, that would lead to an even more thoroughgoing rejection of the original logical positivism. It is the move to this new phase, which in the literature is generally referred to as the 'newer philosophy of science', that we want to look at in more detail in the next section.

2.4 The newer philosophy of science: Thomas Kuhn

Until the mid-1960s the philosophy of science operated for the most part on a normative or prescriptive level. The intention which lay at the root of the discipline was to *prescribe* how science ought to work. On the basis of the philosophical theories that were developed, it was to become clear what should be given the name science and also of course what should not. We saw this both in logical

30 This distinction originates from Hans Reichenbach, *Experience and Prediction*, Chicago 1938, ch. 1 (see also n. 4 concerning Reichenbach, a member of the *Berliner Gruppe*).

positivism with its strongly anti-metaphysical bias and in the critical rationalism of Popper with his focus on the 'problem of demarcation'. By contrast, the contribution of thinkers such as Quine and especially Duhem that we looked at briefly in the last section developed much more from how things worked in practice, in research laboratories etc. In this way it became clear that not only the original verificationism and confirmationism were untenable (because they allowed for too few or too many statements in scientific language), but also that falsificationism was problematic. All in all, the prescriptive approach turned out not to lead to satisfying results.

This development led to a shift in the philosophy of science from a prescriptive to a much more descriptive approach. Instead of prescribing how science should operate, the point of departure more and more became the description of how it does operate in reality. To that end, scholars intentionally started to study the actual history that the sciences (initially the natural sciences in particular) had gone through. Of course, this had also been done before them, but then history had served mostly as a source of illustrations to support one's own theories.[31] From 1970 onwards, this changes, and from then on we see that the aim is to fit theory formation in the philosophy of science as closely as possible to how the historical development of the sciences has in fact taken place. The saying 'Don't think, look!' is heard more and more. Let us stop devising marvelous theories about how science should work, and instead look at how it in fact has worked and does work. That will prevent us from again constructing demarcation criteria etc. which in the final analysis cannot do justice to a reality that is much more recalcitrant and complex than we had imagined.

The name that is inextricably linked with this 'historicizing' of the philosophy of science, and thus with the greatest shift in the history of this discipline, is that of the American historian of science Thomas Kuhn (1922-1996). A million copies of his main work, *The Structure of Scientific Revolutions,* have been sold. This has been translated into more than twenty languages, and has remained one of the ten most cited academic publications for more than thirty years.[32] It is thus difficult to overestimate Kuhn's influence on the overall debate about the signi-

31 Two exceptions to this rule are N. R. Hanson, *Patterns of Discovery*, Cambridge 1958, and Stephen Toulmin, *Foresight and Understanding*, London 1961. Both are, because of their accent on the concrete context within which the practice of science takes place, seen as important precursors of Thomas Kuhn.

32 I have taken these data from Steve Fuller, *Kuhn vs. Popper. The Struggle for the Soul of Science*, Cambridge 2003, 1 (by the way, Fuller's thesis here that Kuhn was the more conservative thinker of the two is not convincing; see below, p. 193, footnote 36).

ficance of science.[33] So we shall first devote quite a lot of detailed attention to Kuhn's work and then finish this section with a short reference to Kuhn's more radical colleague Paul Feyerabend.

Kuhn was a physicist who from early in his career was interested in the history of his discipline, and specialized further in this after completing his PhD thesis. On the basis of detailed historical studies, over time he developed a comprehensive picture of how science in fact operates. He expanded this picture in his monograph on the structure of scientific revolutions.[34] In particular the second, revised and improved edition of this book has become a classic.[35] In this study, Kuhn demonstrates that many of our current ideas about the way in which the natural sciences have developed in the course of history do not make sense, because we have paid too little attention to the numerous fault lines and unexpected twists and turns that have taken place in this process. Anyone who looks closely at this history will discover that science has moved in anything but a neat, uniform line of progression. Rather, there have been cases of stagnation, of rival theories that could not be weighed up against each other in a rational manner, and even of scientific *crises* and full-blown *revolutions*.

One of the causes of why we see so little of the more or less chaotic and unpredictable way in which science has developed, is that in text books such as those used in secondary education, these sorts of fault lines, crises and splits have usually been carefully smoothed over. We only ever read about the final results of all sorts of research, due to attempts as it were to link these in an unbroken chain with those of earlier and later research so giving the impression of an overarching unity. All sorts of dead ends, alternative ideas and protracted debates are at best just touched on, and usually implicitly traced back to obvious

33 In response to this observation in the original Dutch edition, it has been pointed out by Willem B. Drees that in the first three issues of the journal *Philosophy of Science* delivered in 2005, only two out of twenty-five articles contain references to Thomas Kuhn; on the basis of this limited experiment, Drees argues that the influence of Kuhn is over-estimated in the present study. See his review of *Een publieke zaak* in *Nederlands Theologisch Tijdschrift* 60 (2006), 355-356. I do not see, however, how this conclusion follows. Without further research, the near-absence of explicit references to Kuhn in a recent volume of *Philosophy of Science* could with equal justice be explained as a sign of the fact that by now Kuhn's ideas are rather widely accepted instead of still being contested and debated. (Compare the number of references to Einstein in an average present-day journal of physics).

34 Thomas S. Kuhn, *The Structure of Scientific Revolutions*, Chicago 1962.

35 In our account of Kuhn's ideas we will make particular use of this second edition (Chicago 1970). By the way, it is tempting to draw a comparison with Karl Barth here, whose break-through as a leading theologian took place with the second edition of his *Letter to the Romans* (1922).

errors that could have been avoided if more care had been taken or a more scientific approach had been adopted. Thus, the development of science is described almost in the same way as we have sketched that of the philosophy of science here: with the omission of all sorts of details and detours that just take us away from the main road, leaving the impression that the truth can be mapped out step by step ever more precisely along this main road.

Now for didactic reasons, text books more or less have to work this way, Kuhn acknowledges. Pupils and students must always be able to see the main thrust of how a scientific discipline developed. However, tthis does not take away the fact that as a result of this a misleading idea of how things work can become fixed in the minds of whole generations of those involved in science. Ways of doing science that have now been left behind (such as alchemy, the theory of ether that we discussed above, the idea of the four humours etc.) are for example seen as primitive initial stages of the current state of affairs, without it being recognized that at the time people could not rationally determine whether or not they were on the right track. Viewed from the current phase of the practice of science they are considered as unscientific, but at the time they belonged to the accepted forms of normal scientific practice. For example, people had the impression that they could gradually get a better idea of the characteristics of the ether that filled space by using all sorts of empirical research and by forming theories from this. In fact, a radical break was necessary – in this regard Kuhn speaks of a scientific *revolution* (compare the title of his book) – in order to discover that the theory of ether actually led nowhere.

In this context we can also understand the concept to which Kuhn owes a great deal of his reputation and with which his name will forever be associated: that of a scientific *paradigm*. Kuhn used this term in the beginning to cover a wide semantic field. In the confusion that subsequently arose, many of those who appropriated the term gave it their own personal interpretation, and for this reason at the end of his career Kuhn himself preferred not to use it anymore. In essence, however, he used it in two ways. First, in the sense of a sort of standard picture of what may be called good science. We often come across such standard examples in text books, because it is important for the future development of science that students make the method illustrated in such an example their own. This making-it-their-own is not only a matter of learning by heart, but also of understanding and of mastering the specific skills needed in order to operate according to the rules of the game. So, here we are not just talking about verbal knowledge that must be transmitted; we are also dealing here with what Kuhn describes, with an expression of the Hungarian chemist and philosopher of science Michael Polanyi (1891-1976), as the 'tacit dimension' of our knowledge:

a 'silent' aspect that can not be exhaustively expressed in words.[36] At a certain moment, we should 'see' how things work, the penny must drop, we must 'get it'.

In this way people are initiated into the existing scientific tradition and they gradually begin to feel at home in the paradigm in the second, broader meaning that Kuhn ascribed to the term. In this second sense, a paradigm is the all-encompassing set of the partly-conscious, partly-unconscious concepts, opinions, expectations and assumptions by which a scientific community lets itself be governed in its work and thought. To this set belong some very simple definitions of concepts (for example in mechanics of a concept such as 'motion'), but also more complex methodological and theoretical ideas (about how theories are tested, calculations are made etc.), and even deep-rooted metaphysical ideas and scientific values (for example about what the ultimate purpose is of a particular form of scientific practice). We could say: in this broader sense a paradigm has to do with the all-encompassing *perspective* within which things are viewed, with a conceptual framework and sometimes even with a complete world view. Kuhn himself (because of the apparent confusion surrounding the term paradigm) later preferred to refer to this broad and fundamental consensus within a scientific community as a 'disciplinary matrix'.

It is especially with this second, broader meaning that the concept of paradigm has later been appropriated by many people who – rightly or not – applied it to all sorts of areas. Here, however, we shall pay most attention to the way in which Kuhn used the term. According to Kuhn, every area of scientific research begins with a 'pre-paradigmatic period', during which a large range of ideas exist about the type and scope of the particular area of research, and about the goal, methods and meaning of the research to be carried out. There are different schools, each with their own ideas, that are supported by arguments as soundly as possible. Each of these schools disputes the fundamental principles on which the other schools are built. However, when the relevant area of science has 'grown up' (Kuhn speaks of *mature science* in this connection), these different schools manage, in an often difficult process, to reach a sort of consensus about what the particular discipline is concerned with and what is important for good research in it. Often a given idea which inspires and gives direction takes precedence over all others. And this is how a paradigm comes into being, a broad-ranging consensus on all sorts of basic questions. The advantage of this is that people do not have to

36 See esp. Michael Polanyi, *Personal Knowledge. Towards a Post-Critical Philosophy*, London 1958. On Polanyi, who is considered as another important precursor of Kuhn, also see Andy F. Sanders, *Michael Polanyi's Post-Critical Philosophy*, Amsterdam 1988.

challenge ever more the assumptions on which these questions are based, but can rather take them as a definite point of departure for further study.

From this moment on, what may be called 'normal' scientific research takes place, conforming to the rules that are suggested by the paradigm. Normal scientific research is thus by definition 'paradigm-dependent'. For without such an overarching perspective giving us direction, we would not know how we should carry out scientific work. In the ideal situation, there is only one paradigm in a given area of research, shared by all those active in that discipline. The task of these scholars is then to test more closely and develop in all sorts of way the theories that are generated within this paradigm.

Sometimes, however, this turns out to be less simple than it may seem at first sight. In practice, what happens is that one always comes across certain obstinate facts that do not seem to accord with the existing formulation of a given theory. According to Kuhn, each theory has *anomalies*, literally: 'illegalities', that is to say situations in which certain observations are different to what would be expected on the basis of that theory. At this point the question is how such an anomaly can best be explained. And here it turns out that there are always various ways of doing this, but that there is always a certain price to be paid for each of them. In order to overcome the obstinacy of a certain anomaly and to reconcile it elegantly with the rest of the theory, sometimes we have to give up another part of the theory. So, we are as it were continually involved in solving puzzles, and when we complete one section of the puzzle this has a knock-on effect for other parts of the puzzle. In standard science according to Kuhn it is therefore not a matter (as Popper thought) of falsifying one's own theories, for as Duhem and Quine showed and Kuhn confirmed from his own historical research, it is possible to deal with the result of a falsifying experiment in very different ways. It is better to describe the task of science as the solving of puzzles that arise in the course of research.[37] In solving such puzzles, we should as far as possible apply the rules that exist within the reigning paradigm.

But sometimes certain anomalies turn out to be very resistant to attempts to incorporate them into existing theories. Whatever is tried, attempts to deliver a generally satisfactory and convincing explanation of the anomalous observations are unsuccessful. Moreover, sometimes the number of anomalies increases rather than decreases as research continues within a certain paradigm. In such circumstances some scientists may get the uncomfortable feeling that there is something

37 Kuhn intentionally used the somewhat playful term 'puzzle-solving' for the task of normal science in order to convey the idea that the problems to be solved are not (in the first instance at least) life-threatening for the paradigm in question. See Kuhn, *Structure of Scientific Revolutions*, 35-43 (esp. 36f.).

structurally wrong with the paradigm as it is. Is there perhaps somewhere a fundamental assumption that will turn out to be incorrect? That would have big consequences for the whole field of research, and possibly cast doubt upon all sorts of supposedly definitive research results. In this way a field of research can move towards a paradigmatic *crisis*.

A favourite example of Kuhn's is the Ptolemaic world view, according to which the sun is supposed to move around the earth. Since classical antiquity this picture of the world was elaborated on in numerous ways, all sorts of puzzles were solved, ad hoc hypotheses (which were often not very elegant) were proposed for troublesome anomalies, etc. But there were still some observations which could not be explained, and people tied themselves in ever more intricate knots (in the case of the Ptolemaic world view for example they had to assume ever more 'epicycles' in the orbits of the planets) in order to do as much justice as possible to these phenomena. In this way, the Ptolemaic picture of the world gradually reached a crisis point: it looked more and more like a house that had been altered so much over the course of time that it took on the appearance of a heap of rubble and could not be inhabited anymore.

Now in the sixteenth century, it was Nicholas Copernicus (1473-1543) who argued that all the pieces of the puzzle would fall into place more easily if we started from a heliocentric picture of the world, in which the earth moves round the sun rather than vice versa. In proposing this alternative, Copenicus actually drew on an old and subsequently abandoned idea of Aristarchus (3rd century BC). Whether Copernicus was right, however, was far from clear. In contrast to what is often thought to be the case, it was not at all evident in the first instance that the Copernican picture of the world was superior; but when the German astronomer Kepler managed to improve on all sorts of flaws in Copernicus' hypothesis, the heliocentric system gradually gained the upper hand. The consequences of this were enormous, for in the wake of the geocentric picture of the world, all sorts of other ideas based on the thinking of Aristotle were swiftly rejected. In short, this crisis was finally resolved by the so-called 'Copernican Revolution', which led to a paradigm shift.

Just as in this case, when such a shift occurs there is usually an initial period during which several paradigms function side by side, without the supporters of any of them being able to convince the others of the merits of their own paradigm. In such a situation there are no firm rules to indicate how such a conflict can be resolved. The rules that do exist belong after all to a paradigm whose right to exist is now being challenged. In practice it is extremely difficult for scientists who work within paradigms that are at odds with each other to understand each other very well. This is the case because within both paradigms (for the sake of convenience let's just limit ourselves to a situation in which *two* paradigms are fighting for preeminence) the same phenomena are viewed in very different ways. As a result, concepts are defined differently, different methodologies are employed, different interpretations are seen as natural, etc. In this regard Kuhn spoke of the *incommensurability* of different paradigms in one and the same field of research. In other words, they do not have any common standard of measurement (Latin: *mensura*) by means of which we can judge the pros and cons of each.

In this connection, it should be realized that it's not only a question of the lack of a standard of measurement or criterion by which we can determine how the problems that arise can best be solved. It is also the case that there is no agreement as to what the problems in fact are! This arises because the observed facts are situated and interpreted in two totally different conceptual frameworks. Those who use rational arguments to try to convince others of the correctness or otherwise of these paradigms fail, because what is considered to be a rational argument is in fact in part determined by the paradigm within which one lives and works. The only way for a scientist to change paradigms is, according to Kuhn, to undergo a sort of conversion: at a certain moment a light-bulb goes on and in a sort of 'aha moment' one discovers that the research data can be much better explained when a completely different theoretical tack is taken.

At one point Kuhn gives an autobiographical example of how such a sudden turn around can happen. As someone trained in Newtonian physics and mechanics, when he was a PhD student Kuhn tried to get a handle on the physics and mechanics of Aristotle. At first he came across so many ideas of Aristotle that were clearly wrong that he was astonished how the Greek philosopher's system had been able to keep going for so many centuries. Had no-one before Newton noticed at all what sort of fundamental misconceptions there are in Aristotle's *Physics*? This was until Kuhn suddenly realized at a certain moment, that he should not conceive of Aristotle's system as a sort of defective forerunner of Newton's, but rather as a world in itself with a whole set of idiosyncratic assumptions, definitions and interpretations. Such an elementary concept as 'movement' for example turned out to mean something completely different for Aristotle than it did for Newton. Aristotle did not consider 'movement' as the measurable displacement of an object through space, but rather as the development of an object in the direction of an 'inbuilt' goal. So, Aristotle was literally talking about other things than Newton was. When Kuhn put himself in Aristotle's place, he discovered that from that perspective there were a lot less obvious faults in this physics than he had initially thought to be the case. 'I still saw problems in his physics, but they were not that conspicuous, and few of them could be regarded as straightforward mistakes'.[38]

38 See the interview with Kuhn in G. Borradori, *The American Philosopher*, Chicago 1994, 156.

In this regard we might think of the well-known picture of the duck (see below), which if we look at it in a different way, can equally well be seen as a rabbit (or it is a rabbit, which can just as well be seen as a duck). You have to 'see' the other image, in a split second; there must be a 'Gestalt switch', a change of perspective that you cannot persuade someone else to see simply by talking about it. And this, according to Kuhn, is how it works at crucial moments in the development of science. People have to be convinced – suddenly or little by little, just as it can happen with a religious conversion as well – of the superiority of the alternative perspective. In such a change, people often have to let go of many of the convictions and assumptions to which they have been very attached and of which they have been certain for a long time. In fact, Kuhn, in one of his most controversial statements, tells us that the whole world changes whenever we make a paradigm shift.[39]

For that reason Kuhn was not particularly optimistic about the frequency with which such 'conversions' actually take place. Usually a certain paradigm dies out not through this sort of move from one paradigm to another, but rather because at a certain moment all the supporters of a certain old (or even new!) paradigm simply pass away, and they have no more disciples who are willing to follow in their footsteps. In this way, the group of scientists active in the relevant field – Kuhn speaks of the 'scientific community' – gradually moves to the alternative paradigm.

All in all, it remains extremely difficult to indicate why a certain paradigm shift happens at a certain moment. What is clear, though, is that the reason for a shift does not lie in the simple and straightforward falsification of a certain theory. We cannot even say that the new paradigm accords more with all the facts or all the observations. For example, this was certainly not the case in the first instance with Copernicus' system. Although Copernicus' model provided an

39 Kuhn, *The Structure of Scientific Revolutions*, chapter 10 ('Revolutions as Changes of World View'). 'In so far as their only recourse to that world [of their research-engagement, GvdB] is through what they see and do, we may want to say that after a revolution scientists are responding to a different world' (111); cf. 137: '(...) after a revolution scientists work in a different world'.

elegant and satisfactory explanation for important anomalies in the Ptolemaic system, it also brought with it new problems. Only after a difficult and to a great extent unclear process did the Copernican world view emerge as the winner in this battle of ideas, because more and more people became convinced that it was right. But was its victory then pure chance? Could the historical process in fact just as easily have turned out differently? If it had, then the way in which paradigms replace one another and in which science develops would be completely random.

This is a view that is in fact often ascribed to Kuhn. Sometimes he is even attacked for it in reproachful terms: with his full-frontal assault on the prevailing philosophy of science he could well have brought the good name of the natural sciences into disrepute by suggesting that they are the play thing of all sorts of irrational historical and social processes. For if the development of science is not determined by rational argumentation, then at any given moment it comes down to which paradigm is the market-leader and which one has the most popular supporters. These supporters often find themselves closest to the centres of political power, and so know how to obtain the most funds for research in order to corroborate the position of their preferred paradigm. By attaching so much value to this sort of social and political factors, Kuhn is said to have revealed himself as a 'relativist', who as some sort of recalcitrant spoilsport refuses to give his support to the victorious progress of the natural sciences which is obvious to everyone else. Some even think that Kuhn was perhaps intent on fouling his own intellectual nest, just as former believers sometimes cannot stop themselves from having a go at the faith with which they were once brought up…

For Kuhn's radical disciple Paul Feyerabend, such a view of things may perhaps hold some truth.[40] Kuhn himself, however, never endorsed it, and rightly so. For him the only thing that was important was to inquire as accurately as possible how things work in practice. In this connection, he did not support an unbridled irrationalism. In fact, he did consider it possible after all to formulate certain criteria which could be used to compare different paradigms with each other. Such criteria are mostly related to the problem-solving capacity of paradigms. For example, does the new paradigm solve important problems which remained stubborn anomalies in the old one? In addition, does it avoid the occurrence of large or many gaps in our understanding which did not exist in the old paradigm? Does the new paradigm cover more ground than the old one (as, for example, Einstein's

40 Compare his attack on modern natural science as 'that most recent, most aggressive, and most dogmatic religious institution'. P. K. Feyerabend, *Against Method*, London 1975, 15. According to Feyerabend the Copernican revolution, for example, succeeded above all because its supporters used smarter propaganda ….

theory of relativity covered more ground than Newton's mechanics)? Can it predict certain phenomena that were not foreseen within the old paradigm? Does it make more precise calculations possible, and is it more coherent and thus simpler than its competitor?[41]

Kuhn also knew, however, that there were no simple or objective answers to this sort of question. For clearly, opinions can differ over the relative importance of each of these criteria, and in addition one scientist may make a different judgement than others over the extent to which the paradigms satisfy each of them. This also happens because usually there are fundamental scientific values underlying this sort of criteria: questions such as what is considered to be really important and definitive and what is deemed 'sacred' in the practice of the discipline? Here there are no more instructions to hand out or rules to imagine which might lead to uniformity. We just have to wait to see which paradigm turns out to be the most convincing, and in the final analysis it is up to the relevant scientific community to give the answer (those outside the community are usually unable to grasp and evaluate the various considerations).

The new, strongly historically rooted approach to the philosophy of science that Thomas Kuhn championed has been welcomed by many. Even a dyed-in-the-wool logical positivist such as Rudolf Carnap wrote words praising *The Structure of Scientific Revolutions*, because of the sterling research work that underpinned the book. In spite of all sorts of criticism that of course also came its way, as time went on Kuhn's work increasingly turned out to have a lasting influence on the discipline. It is not overstating the case to say that Kuhn's paradigm theory *itself* became a new paradigm. It is a generally-held view that we can speak of a pre- and a post-Kuhnian phase in the philosophy of science. Depending on exactly where one puts the accent, one can speak of a historicising, sociological, or even empirical shift since Kuhn. The attempt to secure a clearly delineated area of pure scientific knowledge in the middle of a sea of metaphysical and ideological superstition has on the whole been replaced by an attempt to delineate the historical and sociological context within which science is empirically (i.e. actually) developing, thus leading to a contextualist instead of a formalist approach.[42]

41 For an account of scientific progress along such lines, see Craig Dilworth, *Scientific Progress. A Study Concerning the Nature of the Relation Between Successive Scientific Theories*, fourth edition, Berlin 2008. On Dilworth's conception, 'scientific theories are seen not to be entities of the sort which are either true or false, but to be structures which are more or less applicable depending on the results of certain measurements' (2f.). In this connection, Dilworth talks about the 'relative acceptability' of scientific theories as determined by criteria such as scope, accuracy and simplicity of the theories (85-88).

42 The term 'contextualism' (as distinct from the former 'formalism') is to be found in e.g. Philip Clayton, *Explanation from Physics to Theology*, London 1989, 18-58.

Whilst Kuhn set off in this direction by especially focussing on the natural sciences, the French philosopher Michel Foucault (1926-1984), quite independently and from a totally different background, came to remarkably similar conclusions for the human and social sciences. Using historical examples, Foucault showed that these (too) have developed with many interruptions, radical discontinuities and phases which fundamentally differ from one another – phases which can each be identified by their own deep knowledge structure, and which he therefore called *épistèmès*, using a term derived from the Greek word for 'knowledge'. So, for example, before 1800 according to Foucault the knowledge structure of that time did not allow for the formulation of theories about work or language.

All this does not mean though that the subsequent development of the philosophy of science as a whole naturally followed Kuhn's paradigm theory. On the one hand after Kuhn thinkers came along who believed that there should be much more fundamental criticism of the claims of modern science. In this regard Feyerabend, whom we mentioned earlier, argued for a 'methodological anarchism' in the practice of science, because in his opinion the strict following of methodological rules only leads to an undesirable uniformity. Real renewal only occurs when people dare to move away from well-known patterns. But instead of consciously doing this, people always attempt to suppress everything that departs from the current paradigm. This being so, Feyerabend criticized the way in which modern science has been placed on a pedestal in modern society. According to him, its assertions are in no way as unassailable as many would like to think and the respect for science takes on a completely unjustified mythico-religious form. In fact, knowledge and progress are not the words which are the key to understanding modern science. Rather, the central category is power: it is as an important social institution that science is particularly keen to maintain and strengthen its position in society. In this way, the critical efforts of Kuhn were still being radicalized by Feyerabend[43].

On the other hand, however, other philosophers of science who published after (or next to) Kuhn tried to bring important qualifications to his ideas, in the hope that they would thereby be able to 'save' the rationality of the scientific project in a more convincing way than could be done by means of Kuhn's paradigm theory. Had Kuhn in fact not gone too far in his criticism of logical positivism and the effect it had on the prevailing picture of science? One of the qualifications proposed is connected with the question that especially preoccupies us in this part of our study, and to which we now return: in the history of the (natural) sciences, can we talk of continual progress? Or is this a myth, too, which cannot stand up to a

43 In section 4.4 we shall take a closer look at the ideas of Feyerabend. Here we limit ourselves to this short observation.

close historical study of how science has in fact developed? We shall look at this question now in the final section of this chapter.

2.5 Scientific progress? The methodology of Imre Lakatos

Thomas Kuhn explicitly discussed the relationship between the terms 'science' and 'progress'. First of all, he points out that in normal scientific practice we can certainly speak of progress. From within the prevailing paradigm people can solve all sorts of problems by formulating appropriate (auxiliary-) hypotheses and theories, refining research results etc. and in doing so they gradually come to unlock the possibilities that the paradigm offers. And here we can certainly talk of progress. This progress is only demonstrable, however, in as far as people share the premises and perspectives of the relevant paradigm. This kind of progress is not essentially different from the progress which is possible in other spheres. Kuhn gives as an example the progress that the theologian can achieve in the explanation of dogma.[44] This is of course possible, but it only counts as progress in a meaningful way for those who share the (in this case religious) outlook of the theologian. In other words: only within certain communities or schools of thought is progress evident and in a certain way even guaranteed, for clearly scholars are always standing on the shoulders of their forerunners.

More interesting, though, is the question as to whether progress can also be demonstrated when there are competing paradigms. Can we say that in the way in which paradigms replace one another in the course of history, progress is in fact achieved? Kuhn described the development of science as being to a large extent cyclical: it goes from a pre-paradigmatic period through the normal practice of science to a crisis, in which the internal contradictions of the pre-paradigmatic phase return. A scientific revolution is then necessary to usher in a new paradigm, etc. The question now is, are we talking about movement in a circle here or rather about movement in an upward spiral? According to Kuhn this is an extraordinarily tricky question to answer. For why should scientific revolutions by definition lead to progress? It is clear that this is how things have to be viewed by those who have experienced and supported a particular revolution, for they think from within the new paradigm and can do nothing other than see this as an improvement on the old one. They will also generally take care to ensure that this picture is presented in the text books which are written from the perspective of the new paradigm, whilst publications written from the perspective of an earlier paradigm are discarded.

44 Kuhn, *Structure of Scientific Revolutions*, 162.

Of course there is a clear reason for this: the new paradigm is chosen precisely because it has managed to resolve more problems than the old one. And not only could a great majority of the old solutions be carried over, as it were, into the new paradigm, but in addition certain stubborn anomalies could also be overcome. This is an important indicator of progress. At the same time, however, we have seen that this sort of consideration does not constitute an unambiguous criterion for paradigm choice. For although there can be good reasons for the choice of a certain paradigm, they are not logically compelling. This is a result of the fact that we can weigh up the merits of individual problem solving strategies in different ways. And whenever paradigms are incommensurable, then it is in fact impossible to speak of progress, because that concept of course presupposes that we have a yardstick with which to compare each of them and to measure one against the other.

Furthermore, we can ask whether in the succession of paradigms that follow on from one another, old ideas which have long been superseded might not in fact gain a new currency. This is a possibility that we should not reject out of hand. We have already seen that on one essential point, the Copernican system referred back to the old idea posited by Aristarchus that the sun does not revolve around the earth, but rather the earth revolves around the sun. Within the Ptolemaic picture of the world, Aristarchus was considered completely out-of-date, and no-one thought anyone would ever come down on his side again.[45] Of course this is not to say that the break-through of the Copernican system implied a return to an older, long forgotten paradigm.[46] Nevertheless, a brilliant intuition of Aristarchus which was not taken seriously for many centuries made an unexpected comeback and was decisively vindicated by the Copernican system.

The conclusion that Kuhn eventually comes to is that we can talk of progress in the sense of the ability of paradigms which follow on from one another to solve problems, but not necessarily in the sense of increasing our knowledge of reality. On a crucial point, Aristarchus in fact stood closer to the truth (i.e. had more knowledge of reality) than Ptolemy (we can say now), although in the Middle Ages, the ability to solve problems of the former's ideas was seen as that

45 It should be acknowledged, however, that the heliocentric system which Copernicus put forward was developed within a different conceptual framework from that of Aristarchus. Compare Kuhn, *Structure of Scientific Revolutions*, 75.

46 Cf. Marcel Sarot, *De goddeloosheid van de wetenschap*, Zoetermeer 2006, 123. Sarot rightly argues that there is no historical evidence for such comebacks of full-blown paradigms that previously had become obsolete; however, no doubt there *is* historical evidence for the fact, that important notions or ideas or convictions that were discarded for a long time may be rehabilitated later on within a new scientific context or paradigm.

much less. All in all this is of course a remarkable fact. For could it not turn out to be the case in the future for example that as a result of continued scientific research the old Judaeo-Christian ideas on the origin of life such as creation and design are much closer to the truth in crucial ways than neo-darwinian concepts like chance and coincidence? But if that does turn out to be the case, can we then still speak of progress? There is only progress, according to Kuhn, in the capacity to solve problems. In this regard, things have continued to improve in the natural sciences. Einstein represents a clear improvement in relation to Newton, and Newton in relation to Aristotle, because if we draw a line from Aristotle to Newton to Einstein, then more and more problems have been solved satisfactorily. But in certain important respects which touch on ontology, Einstein's theory of relativity has more in common with Aristotle than with Newton![47] In short, according to Kuhn, it is not true that science succeeds in gradually filling in more and more blanks in our knowledge of reality.[48] In fact, we can say nothing meaningful about the extent to which it does this.

It will be clear that on the basis of this conclusion at the end of his book, Kuhn is seen by many as a relativist in the philosophy of science. Those philosophers of science inspired by Popper sought (and still seek) opportunities to counter the threat of this relativism. One of the most well-known and most fully developed attempts to do this comes from a disciple of Popper who came from a Hungarian-Jewish background, *Imre Lakatos* (1922-1974). Lakatos was convinced of the structural shortcomings that attach to Popper's falsificationism. This is clear from the wonderful example that he gives of the way in which natural scientists always manage to avoid falsification of their theories. The example concerns a fictive case of 'planetary misbehaviour' and is best described using Lakatos' own words:

> A physicist of the pre-Einsteinian era takes Newton's mechanics and his law of gravitation, (*N*), the accepted initial conditions, *I*, and calculates, with their help, the path of a newly discovered small planet, *p*. But the planet deviates from the calculated path. Does our Newtonian physicist consider that the deviation was forbidden by Newton's theory and therefore that, once established, it refutes the theory *N*? No. He suggests that there must be a hitherto unknown planet *p'* which perturbs the path of *p*. He calculates the mass, orbit etc. of this hypothetical planet and then asks an experimental astronomer to test his hypothesis. The planet *p'* is so small that even the biggest available telescopes cannot possibly observe it: the experimental astronomer applies for a research grant to build yet a bigger one. In three years' time the new telescope is ready. Were the unknown planet *p'* to be discovered, it would be hailed as a

47 Kuhn, *Structure of Scientific Revolutions*, 207 (from the 'Postscript – 1969').

48 One of the reasons why Kuhn criticizes this view of things concerns the so-called theory-laden nature of human perception. For didactic reasons we do not discuss this point until Chapter 4. There it will also become clear why Kuhn's conclusion, although it goes against our intuitions, is so difficult to refute.

> new victory of Newtonian science. But, it is not. Does our scientist abandon Newton's theory and his idea of the perturbing planet? No. He suggests that a clould of cosmic dust hides the planet from us. He calculates the location and properties of this cloud and asks for a research grant to send up a satellite to test his calculations. Were the satellite's instruments (possibly new ones, based on a little-tested theory) to record the existence of the conjectural cloud, the result would be hailed as a outstanding victory for Newtonian science. But the cloud is not found. Does our scientist abandon Newton's theory, together with the idea of the perturbing planet and the idea of the cloud which hides it? No. He suggests that there is some magnetic field in that region of the universe which disturbed the instruments of the satellite. A new satellite is send up. Were the magnetic field to be found, Newtonians would celebrate a sensational victory. But it is not. Is this regarded as a refutation of Newtonian science? No. Either yet another ingenious auxiliary hypothesis is proposed, or ... the whole story is buried in the dusty volumes of periodicals and the story never mentioned again.[49]

So it turns out in practice to be almost impossible to falsify completely an established set of theories such as those of Newton. Whatever happens, we may always try to find a way round. There is as it were an unsolvable problem of adaptations: In principle we can carry on forever bringing forward ad hoc adaptations or adjustments to the overall web of our theories, in order to immunize our best-loved theories against criticism.

It is important to see, however, that Lakatos does not just use this example in order to ridicule the working methods of this imaginary physicist. On the contrary, what this physicist did was completely rational, given the state of affairs in his discipline at the time. For clearly, in the past the theories of Newton were confirmed so many times and were capable of resolving so many problems that much more than a single anomaly would have been necessary to cause them to be rejected, however much that anomaly persisted. In any case, there should first of all be an encompassing *alternative* to which we can move. In this instance, an alternative did not arise until the work of Einstein, and only then the deviating orbit with which this whole story began could be satisfactorily explained. Until this point, the stubborn dogmatism with which 'our physicist' held to his Newtonian set of ideas could be seen as entirely rational.

So, Lakatos understood well that a 'naïve falsificationism' such as that which, according to him, we find with Popper, does not work. Nature does not call out such a resounding 'no' as Popper thought. But, Lakatos does not go along with the idea that there should therefore be no rules whatsoever that tell us when we should hold on to a theory (or set of theories) and when we should

49 This example comes from Lakatos' best known publication (in his case not a book but an extensive article): Imre Lakatos, 'Falsification and the Methodology of Scientific Research Programmes', in: I. Lakatos and A. Musgrave (eds.), *Criticism and the Growth of Knowledge*, Cambridge 1970, 100f.

reject it (or them). If the path that scientists take cannot be rationally justified (not even in retrospective), then Kuhn could be right in saying that no-one can control how scientists make their theory choices, and that sociological factors therefore might in fact play a decisive role in making them. Indeed, Lakatos reproached Kuhn for suggesting that the development of science was not governed by the search for truth, but completely by the struggle for power.

This reproach was somewhat over-simplified, but led Lakatos into a serious attempt to deal with the relativistic threat that he saw rearing its ugly head with Kuhn. Lakatos himself wanted a sort of middle way between the old empiricist philosophy of science on the one hand and the newer philosophy of science on the other, in the hope of thereby being able to safeguard what for him were some of the essential elements of the first from the incisive critique of the second. And so he searched for a set of rules which could rationally justify the choice for or against a certain theory,[50] or in one word: he looked for a *methodology*, a *logos* – i.e. a rational exposition – of the *methodos*, the way that must be followed in order to achieve real progress in scientific research. Lakatos saw that such a methodological justification is required in order to do away with the impression that there is a complete arbitrariness in the development of the sciences. To put it in other words, he looked for a way of rescuing the distinction between the (contingent) context of discovery and the (rational) context of justification of science from the jaws of the newer philosophy of science.

Central to the methodology that Lakatos puts forward is the category of a 'research programme'.[51] By this Lakatos means a complex network of inter-connected theories, auxiliary hypotheses, generally accepted empirical observations etc. The idea here is that these research programmes continue to generate new independent theories which can be researched on their merits. In this regard, Lakatos' research programmes may be compared with Kuhn's paradigms, although they can be more limited in scope. An objection to Kuhn's concentration on the concept of paradigm is that attention here is very much focussed on the few all-encompassing shifts that we encounter in the history of science, such as those from the Ptolemaic to the Copernican world view, from the belief in creation-within-six-days to Darwin's theory of evolution, from Newton to Einstein etc. Such fundamental shifts (with all their implications for how large groups of

50 There has been a flood of recent literature about terms such as 'rational', or 'rationality' as there has been about what we should understand by the 'justification' of knowledge claims. Cf. e.g., to mention only one title, Mikael Stenmark, *Rationality in Science, Religion, and Everyday Life*, Notre Dame 1995. Here we do not use these terms in any sort of technical sense, but in their 'common sense' meaning.

51 Lakatos formulated his methodology of scientific research programmes for the first time in 1968.

people experience reality) are however few in number. They may capture the imagination, but the average scientist often has hardly anything to do with them in his or her own work. On the other hand, it could be said that scientists usually experience their work as taking place within a certain research tradition or research programme. Such programmes can quite happily stand side by side for a long period in a given area of research, and are less 'incommensurable' than Kuhn's paradigms.

The question now for Lakatos is how such competing research programmes can be compared with each other and evaluated in a meaningful way. Indeed, it is in fact programmes such as these, so somewhat larger theoretical-empirical complexes, that are compared with and weighed against one another, and not (as Popper thought) individual theories. The example we discussed above of the imaginary physicist who attempted to explain the unexpected planetary behaviour shows clearly that a theory can never in fact be judged on its own, in isolation from the larger whole of which it is a part. In this sense Lakatos has accepted the holistic approach of Quine,[52] although he 'stretches' this less than Kuhn. For whereas according to Kuhn the co-existence of two rival paradigms is an exception which signifies a crisis, for Lakatos it is the rule that research programmes will compete with one another. Lakatos even supports the idea of his friend Feyerabend that in principle we must strive after pluriformity in this area! For clearly, as long as there is no alternative to a given research programme, it can never be refuted – as again the example we have used shows. And here Lakatos agrees with Popper that research programmes which cannot be refuted only hinder the progress of science.

Of particular interest is Lakatos' view that research programmes in principle consist of two parts, namely a 'hard core' and a 'protective belt'. This insight is an important refinement and improvement in relation to Kuhn's views.[53] The distinction between a hard core and a protective belt arises because within a research programme two sorts of rules apply: rules that determine what internal changes are not to be taken into consideration when someone looks for solutions to certain problems (these rules form the 'negative heuristic') and rules which suggest what paths you can go down whenever you want to rectify an existing problem or adopt a new practice (together these form the 'positive heuristic'). The first group of rules serves as it were to safeguard certain assumptions which are considered central to a given research programme. These should not be given up out of hand whenever a problem occurs, for

52 See above, section 2.3.
53 Lakatos considered himself much more as someone attempting to refine the falsificationism of his teacher Popper (although the latter did not see the need for this; for a discussion of their relationship, see Theo A. F. Kuipers, 'Poppers filosofie van de natuurwetenschappen', *Wijsgerig Perspectief* 42.2 (2002), 26-28). In practice, however, his views can equally well be compared with those of Kuhn.

they are to a high degree constitutive of the whole programme: they form its *hard core*. If there are problems or if you want to develop the research programme in a certain direction, with the consequence that you have to implement certain changes to it, then the second group of rules indicates how you can do this. Here, you must formulate (or adapt) certain testable auxiliary hypotheses, that together can form a *protective belt* around the hard core. These auxiliary hypotheses are intended to provide a better explanation for existing observation data or to provide room for data which have not yet been assimilated by the research programme.

This last point is particularly important for Lakatos. In contrast to Kuhn, he does not think only in terms of anomalies that must be resolved, but above all in terms of 'novel facts' that must be brought to light. This can of course happen through the resolution of anomalies, but it can just as well happen by unexpected applications or the making of predictions from auxiliary hypotheses that are related to the existing hard core of the research programme. By such novel facts Lakatos does not necessarily mean literally new facts, in the sense of thus far completely unknown states of affairs; it can also be facts which are already known but which are suddenly be seen in a surprising new light, or which turn out to accord precisely with a new theory, or again which can be explained much better (more elegantly, more simply, more consistently etc.) as a result of it. For this reason, Lakatos deliberately speaks not of new but novel facts.[54]

According to Lakatos, the prediction and vindication of such new facts or explanations[55] is to be seen as the primary task of every scientific research programme. And it is here that Lakatos in fact sees the opportunity to separate the wheat from the chaff. His position is this: a research programme is progressive, as long as the theories it consists of continue to generate novel facts, that is as long as they continue to predict facts that could not have been imagined beforehand. And Lakatos distinguishes here between theoretically progressive programmes in which such novel facts are only predicted, and empirically progressive programmes, in which they are also actually confirmed through further research (by withstanding tests which are aimed at their falsification). Whenever the latter succeeds for at least some new facts, then the non-falsified empirical content of the relevant

54 See (also for his favourite examples) I. Lakatos, 'The Methodology of Scientific Research Programmes', in: id., *Philosophical Papers*, Vol. 1, Cambridge 1978, 184f.

55 There is no overall agreement in the current philosophy of science literature as to what exactly an explanation in the natural sciences is. For this discussion, see Wilko van Holten, *Explanation within the Bounds of Religion*, Frankfurt 2003, 20-73, and Michael Strevens, *Depth. An Account of Scientific Explanation*, Cambridge (Mass.) 2008. In line with a leading position (viz. the so-called 'unification model'), we especially use the term here to express the idea of 'making sense of' a given fact: increasing our understanding of a given observation by convincingly placing it within a broader context.

research programme clearly increases. And it is precisely in this type of increase that we find the objective progress of science and of our knowledge of reality.

However, if a certain change in a research programme provides us with no novel facts, then such a change may be seen as a more or less *ad hoc* adaptation, simply inspired by the wish to save the hard core. For in such a situation the truth of the newly proposed (or revised) auxiliary hypothesis cannot be established in an independent way. In such cases, Lakatos speaks of a degenerating programme. If we then have to choose between two competing research programmes, it is possible to make this choice rationally: for we must choose the most progressive and least degenerative programme. In this way there really is something sensible to say about the rationality of theory choice and about the progress of science through the succession of different research traditions.

But let's not jump for joy too soon. For there is an important complication in the question of how to make the choice for or against a given research programme. Lakatos has to acknowledge that it is possible that a certain research programme may degenerate *for a time*, but subsequently bring forth novel facts again – somewhat like a tree that temporarily does not bear fruit but does so again when the time is right. For we cannot exclude the possibility that researchers who are active within a degenerating programme may manage at a certain moment to put matters to rights and get back on track.

An interesting example of this on a micro-level is the so-called theory of Prout from 1815. Reduced to its essentials, this theory, developed by the chemist William Prout, asserted that the atomic weight of all pure chemical elements should be a multiple of the atomic weight of hydrogen (given that all atoms are composed of hydrogen atoms). For years there was no way of confirming this theory. On the contrary, on the basis of empirical tests it turned out time and again that for example the atomic weight of chlorine was 35.5 times that of hydrogen, et cetera. However, at that time there was only one definitive method for testing the atomic weight of elements, and in order to save their theory, the Proutians had to assert that there was something wrong with this method. That seemed to be some sort of desperate ad hoc hypothesis, for there was nothing to suggest that this was the case – until in 1911 a new method was discovered in the light of which the old method did indeed turn out to be inaccurate. When scientists calculated the atomic weight of elements with the help of this new method, the falsifications of Prout's theory became confirmations, so that his theory at once brought to light an amazing range of novel facts. But it took all of ninety-six years (the time between 1815 and 1911) before Prout's theory changed from a degenerating programme into a progressive one![56]

56 For this example, see Herman Koningsveld, *Het verschijnsel wetenschap*, Meppel/Amsterdam 1980, 185f. This example of course operates on the level of a single

For this reason we can understand why Lakatos argued for a certain amount of tolerance in relation to degenerating research programmes. But the question is, where does this tolerance end? On this point Feyerabend offered some serious criticism of Lakatos' methodology. Feyerabend asserted that there is a dilemma here: either we put a definite limit on the amount of time we give a degenerating programme, or we do not do so to be on the safe side. The first of these would put an unjustifiable limit on scientific freedom; for obviously, we can never exclude the possibility that shortly after we have passed the time limit, a degenerating programme will ultimately start to produce results! But the second would imply that in practice Lakatos' methodology cannot work, because we would continually be giving endless credence to degenerating research programmes. In that case, however, the development of science no longer follows strict rules, but it is just as capricious as Feyerabend himself thought. In addition, the notion of objective progress in our scientific knowledge is once more put under great pressure. For everything can always turn out to be very different from what we had previously thought.

Unfortunately, Lakatos did not live long enough to be able to reply to this criticism. But it is clear that his methodology must be at least considerably refined in order to keep going. Moreover, it could be argued that to some extent Lakatos again wanted to prescribe from some sort of ivory tower how things should operate in science without learning sufficient lessons from how it actually does operate. The tension between prescriptivity and descriptivity, which was dissolved so consciously by Kuhn to the benefit of the latter, in any case makes a complete return with Lakatos. Finally, it is true that Lakatos looked for a new, objective methodology, but in his epistemology he continued to concur fully with his teacher, Popper. That means that he as well had to accept the essential fallibility of human knowledge and also had no adequate response to the Duhem-Quine thesis. The example of Prout's theory shows us once more, that a critical experiment (i.e. an experiment that must determine whether a certain theory is correct or not) is not possible, because even in the simplest of situations we cannot determine definitively what part of a theory or which of its underlying premises is refuted by falsification. From this example it is also clear that as a result of the impossibility of such a critical experiment, it cannot be definitively determined whether a research programme can be seen as progressive or not.

theory, not on that of a more complex research programme. But on that level things are no different from a structural perspective; it is just more difficult to give a simple example for that.

For this sort of reason, Lakatos' methodology has found little support.[57] Although others have tried different ways of reaching the goal that Lakatos had in mind, none of them has so far managed to do so in a way which speaks to a wider audience. That said, Nancey Murphy believes that Lakatos' account of what constitutes scientific rationality is still the best that we have.[58] If this is true, then we must conclude that the prospects of being able to speak with justification of scientific progress are bleak. That is, if we understand progress in the sense that Kuhn disputed, namely as the increasing knowledge of reality. In any case, as we come to the end of our journey through the history of the philosophy of science, we have to conclude that the popular image of science as a linear-progressive affair, in which more and more blanks in our knowledge of reality are gradually filled in is incorrect. In practice, things in the scientific project turn out to go in much less of a straight line. The development of science in fact encounters numerous fault-lines, reversals, crises, chaotic situations and long-winded dead-ends. It is without doubt the virtue of the newer philosophy of science that it has opened our eyes to this.

The first pile supporting the standard image of science therefore is, to put it mildly, less secure than is often thought to be the case amongst lay people (and indeed scientists as well). This raises the question about the reliability of the other two piles that we identified in section 2.1. We shall take a closer look at these in the next few chapters.

57 This does not take away the fact that important elements of Lakatos' theory of science still have lasting value, such as the distinction between a hard core and protective belt in research programmes.

58 Nancey Murphy, *Theology in the Age of Scientific Reasoning,* Ithaca 1990, 61. Murphy compares Lakatos in particular with Larry Laudan, *Progress and its Problems,* Berkeley 1977, and argues for the superiority of Lakatos' methodology. Accordingly, she positions herself as a conscious follower of Lakatos. In this connection, it is interesting that she notes Feyerabend's criticism of Lakatos' methodology discussed above, but does not make any attempt to refute it.

3. Multi-coloured reality

3.1 Second characteristic: the natural sciences as text-book example

In the last chapter, we saw that the credentials of an important part of the standard picture of science are less sound than is often recognized. The idea that science simply moves in a steady upwards line turned out to be incorrect. It is without doubt from this idea that science gains part of its powerful image. And this will continue to be the case for as long as scientists manage to create new technical applications from their research. But this image does not in fact accord with the much more fickle way in which scientific research develops in reality. And here it is not at all self-evident that we gain an ever-increasing hold on reality. The newer philosophy of science showed us that our knowledge of reality is strongly determined by paradigms and that it is impossible to compare divergent paradigms with each other from an Archimedean perspective (a 'God's eye point of view', as it has been called). And it transpired that it was not even out of the question for core parts of paradigms which had long been abandoned to be 're-habilitated' again many centuries later. Human knowledge remains forever fallible.

The question we ask in this chapter is how to analyze and evaluate the second element of the standard picture of science that we identified above. Is it indeed the natural sciences that set the trend, so that the other branches of scholarly practice must, as it were, do their best to copy the method of the former as much as possible – at the risk of not being seen as a fully paid-up academic discipline? Or is there in fact more of a variety of scientific methods? We shall consider these questions in the next two sections. We shall demonstrate that there is indeed more variety, and then in sections 3.4 and 3.5 consider the question of how we can account philosophically for this variety.

3.2 A spectrum of sciences

In the previous chapter we saw that in the philosophy of science a crucial shift has taken place from a prescriptive approach to a more descriptive one. The phenomenon of science is now no longer primarily approached from top down, being subject to a raft of a priorily determined demands that it has to meet, but rather from bottom up, with close attention being paid to the problems that scientists have to deal with in practice, and to how they do this. If we follow this approach, then it is natural in the first instance to consider what sort of research is carried out in those institutions considered scholarly centres *par excellence*, viz. the universities. The first thing that strikes us here is without doubt the staggering variety of areas and methods of research that we find in an average medium-sized university. The mere sight of this enormous variety can lead us to ask questions about the dominant position that the natural sciences occupy in the standard picture of science.[1] Clearly, there is a difference here between image and reality. This is without doubt connected to the widespread reverence for the natural sciences that we see across society. Presumably this leads many people, without even thinking about it, to equate all scientific and academic work more or less with the natural sciences.[2]

Countless attempts have been made to bring some order to the multitude of academic disciplines and areas of research. Such divisions as there are do of course have their own significance: they help us to be aware of the underlying relationships between different areas of research, without losing sight of their multiplicity. But each division also has disadvantages, because it is always the case that some relationships are highlighted whilst others are not. Moreover, their boundaries seem to be in a constant state of flux, because inter-disciplinary research often leads in the long run to newer, more or less independent areas (and sometimes even methods) of research.

At the Dutch Organization for Scientific Research for example an eight-fold division of sciences is currently in use. In alphabetical order these are chemical sciences (four sorts of chemistry), earth- and life sciences (with amongst others biological, geological and meteorological research), exact sciences (mathematics, information technology and astronomy), humanities (history, art history, archaeology, linguistics, literary theory, theology, philosophy, and film, music and theatre studies), medical sciences, physics, social and behavioural sciences (under which heading come all areas of social scientific research: psychology, pedagogy, educational theory, business sciences, economic sciences, environmental sciences, social-cultural sciences, law, the science of public administration and communication sciences, political sciences), and technical sciences.[3] It is clear that such a division

1 Translator's note: the reader is reminded of the fact that the term 'science' in this context includes all academic forms of knowledge production (cf. above, chapter 2 n. 2).

2 In this regard, it is interesting that the Dutch organization for Pure Scientific Research (ZWO) dropped the epithet 'pure' when it became the Dutch Organization for Scientific Research (NWO). Here, we see the influence of the shift from the older to the newer philosophy of science: and indeed we do not find it so easy to say what 'pure' science is.

3 This information is taken from the NWO website, www.nwo.nl.

is in part based on practical considerations (existing interdisciplinary research structures etc.), for from the point of view of content astronomy seems closer to physics than to information technology, whilst biological research has at least as much to do with chemical as it does with geological research.

A division based more on substantive differences and similarities in object and method could look something like this:

1. First of all there are the *formal sciences*, which are characteristically based on conceptual systems and not on empirical reality. Those that come to mind first are mathematics, statistics, symbolic logic and, more recently, cognitive artificial intelligence. Because these disciplines have nothing to do with the empirical world, we cannot use empirical proof for them. That said, they are used in many ways in empirically-oriented sciences, and gain much of their importance in society from this (without mathematics it is difficult to build bridges et cetera). Straightaway, though, we meet here the problem that we have with every division, for symbolic logic is in practice often considered to belong with philosophy, one of the humanities (see below).

2. The *natural sciences* study the empirical reality in as far as it is not affected by human behaviour (apart from the experiments in which we do try to ascertain its effects). They explain what is observed in such cases in the empirical reality, by placing it in a wider context using natural laws that have already been established. On the one hand, we can include amongst the natural sciences disciplines aimed at acquiring knowledge such as physics, chemistry, biosciences, astronomy and geology, and on the other hand disciplines with a practical application such as the technical sciences (architecture, materials science etc.) and medical sciences (medicine, pathology, virology etc.). In addition, research on so-called quantum phenomena belongs to the natural sciences, although it must be asked here whether the influence of human behaviour on the research results can really be excluded[4].

3. The *social sciences* have something in common with the natural sciences in that both are empirical in nature. But in contrast to the natural sciences, the social sciences are concerned with human behaviour and the influence of this behaviour on reality. In broad terms, the natural sciences deal with nature whilst the social sciences deal with human beings and society. Amongst the social sciences we find the disciplines that we mentioned above under the heading social and behavioural sciences: psychology, sociology, pedagogy, economic sciences, law, political sciences, study of religion, et cetera. To this list, we must also add linguistics, because language is also a mode of human behaviour. To the extent that 'laws' are to be found in the social sciences, these are of a statistical nature and not, as we usually find in the natural sciences, of a causal nature.

4 For more on this, see section 7.4.

4. *History and literary theory*, which are often incorporated into a Faculty of Arts, likewise study reality as it is shaped by humans (and together with the social sciences we can also refer to these as the cultural sciences), but with specific reference to the past. Literary theory is concerned with the interpretation of texts from the past, whilst history is concerned with the interpretation of events from the past. Both are closely connected to each other: historians can never do their work without the interpretation of documents, literary theorists always need to know about the historical context of the texts which they study. Closely linked to both of these branches is archaeology, which is concerned above all with the study of artefacts from the past other than texts, although it does of course make use of ancient texts in its work. In the forming of historical and literary theories there is never any appeal to (causal or statistical) predictive laws, because in principle these disciplines deal with unique events and texts.

5. Finally, the *humanities* are concerned with questions which in the final analysis cannot be resolved empirically (which does not mean to say that the empirical world does not have an important role to play in them). This basically non-empirical character is what the humanities have in common with the formal sciences. The questions that are of central concern in the humanities, however, are not of a conceptual nature, but of a normative one. They include amongst others disciplines such as philosophy (in particular sub-disciplines such as philosophical anthropology, the philosophy of science, ontology/metaphysics etc.), ethics and theology.[5] Here, we are dealing with questions such as: What is determinative for our existence? How should science be carried out? How should reality be understood as a whole and in its constituent parts? Are there moral obligations which we should fulfil? What is the meaning of our life and of reality in relation to God? Although within these sorts of questions all types of empirical questions are concealed which are dealt with at length by these disciplines, the real questions are ones of meaning. Following a long tradition, these questions as well can be studied in an academic manner (indeed, the university owes its origins to theology).

This classification of the most important categories of academic research is not exhaustive and also not without problems. The framework that we have set out is for example too crude to find a place for complex disciplines which combine approaches from different categories with one another. For example, environmental science combines the research methods of both the natural sciences

5 Translator's note: In many cases the Dutch term for 'humanities', namely 'geesteswetenschappen', is defined more broadly in keeping with its German origins, and as well as philosophy and theology also includes the disciplines of history, literature, art and linguistics. Thus, it is a rather convenient term with which to cover some very disparate entities. The difference we describe here between historico-literary studies and the humanities offers more clarity in this regard.

and the social sciences with each other, and many argue that cosmology combines data from the natural sciences with insights from the humanities (especially philosophy and/or theology). In addition, the natural sciences for example in a certain sense have mathematics as a prerequisite, whilst the reverse is not true. The above classification does however allow us to see clearly how inappropriate it is to – either consciously or unconsciously – derive all scholarly academic work from natural science. There is so much more at stake in the world of academic research. This 'more' can neither prima facie be seen as inferior, nor be easily reduced to the natural sciences.

If we now look at how the philosophy of science has dealt with this diversity, then we see that at first it paid almost exclusive attention to the foundations of the natural sciences, whilst more recently there has been increased reflection on the nature of the social sciences. However, much less attention has been given to the status of disciplines that belong to categories 4 and 5 above.[6] Also in recent studies on the philosophy of the humanities (in the broadest sense of the word) these are hardly dealt with separately at all. This is a pity, for as a result representatives of these disciplines often still tend to consider their own specialisms as second-rate academic disciplines, because of the implicit ideal of a monolithic methodological unity based on the natural sciences (an ideal which even the social sciences at best can only meet in part). But there is hardly any reflection at all on how these individual disciplines relate to each other from a methodological perspective. Presumably this is in part a result of the fact that in current scientific research the prevailing way of disciplinary categorization of the sciences is being challenged more and more, although this categorization still has great importance from a social and educational perspective – a state of affairs which will presumably continue for the time being.[7]

3.3 Natural sciences and humanities: a false dichotomy?

We cannot look any further now into the diverse nature of each of the categories that we list above, nor into their status from the perspective of the philosophy of science, but we can discuss the important question of whether or not the social

6 A nice exception to this rule is P. B. Cliteur et al. (eds.), *Overtuigend bewijs. Over het wetenschappelijke van de niet-exacte wetenschappen*, Amsterdam 1994. This contains the fruits of a series of lectures given at Leiden University which deal with the scientific character of law, literary theory, history, theology, philosophy and women's studies.

7 On this subject, see Michiel Leezenberg & Gerard de Vries, *Wetenschapsfilosofie voor geesteswetenschappen,* Amsterdam 2001, 231-233 ('the status and erosion of disciplines').

sciences have a separate status from the natural sciences. This is the subject of an old discussion which has a great influence on the 'big picture'. If the social sciences have or should have the same methodological character as the natural sciences, then there would only be a few exceptions around the edges, whose status as scientific disciplines should perhaps be questioned indeed. This is of course not true for an exact science such as mathematics, but it would be true for the arts and the humanities.

Conversely it is perhaps the case that if the social sciences do have a methodological perspective which differs structurally from the natural sciences, it is plausible to see the arts and humanities as disciplines which also have their own, but no less scientific, methods. In that case, it would clearly be wrong to demand that a discipline such as theology mirrors criteria which are applicable to the natural sciences, without taking into account the specific nature of 'intermediate' disciplines such as social sciences and the arts. So, an important question to ask is whether the social sciences can and should reflect the 'strong' natural sciences from a methodological perspective, or whether they are of a different order.

This discussion has been going on since *Wilhelm Dilthey* (1833-1911) argued at the end of the nineteenth century that there was a fundamental difference between on the one hand sciences which have nature as their object, and on the other hand those which study culture, or the products of the human mind from the past and the present. This latter category, which Dilthey was the first to call the *Geisteswissenschaften*, is characterized by a *verstehende* method in contrast to the *erklärende* method of the natural sciences. It is difficult to translate this word *verstehen* into English, as it can mean 'to understand', 'to hear' and 'to interpret'. In the beginning Dilthey used it primarily in the sense of 'to understand'. Products of cultural endeavour (such as a system of law, a work of art, or even a moral system, a religion etc.) always have some specific meaning as a result of the fact that the human mind expresses certain intentions, ideals and/or values through them. In trying to find this meaning, we do not get very far by continually asking what the causal explanation of the relevant products may be (this only provides us with the knowledge of how they were made or came about). Much more relevant knowledge comes about when we learn to understand, or grasp the meaning of, these products, and we do this by trying as far as possible to inhabit the mind of other people, for example of the artist who has created a sculpture.

Thus, in the first instance Dilthey conceived of his *verstehende* method in strongly psychological terms: what matters is the personal ability of an academic in the humanities to empathize. Through a process of introspection this person must as it were try to feel the experience of others. Later, however, Dilthey substituted this approach with a much more hermeneutical and non-psychological formulation of the *verstehende* method. In retrospect, this was an excellent move, for this hermeneutical method should have a much longer life-expectancy. It is not so much a question here of trying subjectively to have the experience of another (and in any case how far this is possible is of course something which is difficult to measure accurately), but more a matter of *interpreting* as well as possible the actions of someone else, taking into account the various formative influences on them. These influences can for example be of a cultural or biographical nature, or depend on someone's view of life. In fact it is true to say that anything that can help to provide insight into the meaning of human actions (or the products of these actions) is to be taken into account. And in each case we always need to look at a greater whole in order to gain insight into the individual artefact: a sculpture can only be understood within the context of the entire oeuvre of the particular artist, and again this oeuvre represents a certain style, which itself can be traced back to certain roots (for example, the desire to distinguish itself from a previous tradition) et cetera.[8]

Given that these sorts of influences usually belong to the public domain, the hermeneutical approach should not be considered as a subjective method, but as an intersubjective one. For clearly, we can meaningfully exchange ideas about the way in which all sorts of factors that contribute to a correct interpretation should be weighed up. Dilthey therefore saw his *verstehende* method as one which, to be sure, was quite different from the explicative method of the natural sciences, but no less scientific for that. He argues that the humanities (in the broad sense of the term, so including for example history; sociology was at this point still not a separate discipline) lead just as much in their own way to knowledge as do the natural sciences, and therefore deserve an equal place in the universities.

Dilthey's ideas were further developed and corrected by the neo-Kantian Heinrich Rickert (1863-1936), who, instead of humanities, preferred to speak of cultural sciences. These are to be distinguished from the natural sciences by their *idiographic* approach to reality. This literally means that cultural sciences are intended to describe the particular (Greek: *to ídion*), for example the particular meaning of this canvas, or of that musical composition. By contrast the natural sciences approach reality *nomothetically*, that is literally by putting forward laws (Greek: *nomoi*). In doing so, they abstract from the particular (what distinguishes this canvas or that composition from all others), and seek the general: which laws apply for all canvases, or for all musical sound? The natural sciences, according

8 In passing, we note that knowledge of these greater wholes is of course only possible by taking account of separate products of the mind. Accordingly, taking his lead from the theologian Friedrich Schleiermacher, Dilthey spoke in this connection of a 'hermeneutical circle', whereby in the process of interpretation over and over again we go from the whole to the parts and back to the whole.

to Rickert, therefore employ a generalizing method, and the cultural sciences an individualizing one – but both may exist and have every right to do so.

Understandably, strong opposition to the idea that there should be more than one scientific method came from the neo-positivistic camp. In particular Carl G. Hempel (1905-1997) tried to show that only one type of scientific method is possible, and that the humanities cannot achieve results by means of that method. Following a path taken previously by Popper, he described this method as the search for *explanations*.[9] According to Hempel, science is concerned with the explanation of phenomena, and a sound explanation has the form of a certain type of argument. One of the premises in such an argument consists of one or more universal laws, and the other consists of one or more individual, actual situations. Now ideally, the observation which needs explaining can be derived deductively from both premises. In this case we can speak of a deductive-nomological model of explanation. We can illustrate this model by using a standard example (in which P stands for premise and C for conclusion):

P1: Copper expands on heating (universal law).
P2: This particular length of copper wire is directly exposed to heat (actual situation).
C: This copper wire is expanding (the observation requiring explanation).

In this way we have explained the expansion of this particular length of copper wire by subjecting this event to a universal law of nature. The event is as it were 'covered' by the relevant law, for which reason we also talk of a 'covering law model'.[10]

The strength of this model is that, because of the explicitly deductive relationship between premises and conclusion, it does not only have an explanatory power but also a predictive one. As long as we know that copper expands on heating, and also that a given length of copper wire is exposed to heat, then we can predict with certainty that, given certain standing conditions, the copper wire in

9 See Carl G. Hempel and Paul Oppenheim, 'Studies in the Logic of Explanation', *Philosophy of Science 15* (1948), 137-175. See also the clear account given by Wilko van Holten, *Explanation*, 38-46, and more generally Wesley Salmon, 'Four Decades of Scientific Explanation', in: Philip Kitcher and Wesley C. Salmon (eds.), *Scientific Explanation,* (Minnesota Studies in the Philosophy of Science, Vol. 13), Minneapolis 1989, 3-219.

10 A second, weaker 'covering law' model of explanation that Hempel later (out of necessity) differentiated is the so-called inductive-statistical model. Here, it is a matter of probable explanations on the basis of statistical laws (such as for example if someone comes into contact with someone else who has an infectious illness, there is a certain percentage chance that they themselves will become infected).

question will expand. We can in fact say that the heating of the copper wire is the *cause* of its expansion. This is not the result of some sort of secret property or tendency of the copper wire (for example 'expansion causing'), but rather of the fact that the law described in P1 obtains. The deductive-nomological model of explanation is in that sense a causal model.

Does such a model also have currency within social scientific research? Or to put it another way: can human activities be explained causally?[11] According to Hempel in principle they can. The Dutch philosopher Rein de Wilde provides a nice example of how such a 'Hempelian' example could look in the social sciences. He points to a debate at the end of the nineteenth century concerning the ideas of Emile Durkheim about a sociology which follows the same pattern as science.[12] According to Durkheim we gain much more insight into society if, in studying social processes, we systematically turn to the natural sciences, rather than content ourselves with a knowledge of the works of, for example, great men of letters. But his plea for a scientific sociology generated fierce criticism, above all from literary circles. How can we explain this criticism? One idea is that the literary critics felt that their interests were under threat. If Durkheim's ideas were to gain ground, that would be at the expense of the place of faculties of literature in the university world. If we put this explanation into deductive-nomological form, it looks like this:

P1: Whenever groups of people feel that their interests are under threat from particular ideas, they will always (if they get the chance, and do not become ill etc.) polemicize against those ideas (universal law).
P2: A group of literary intellectuals at the end of the 19th century felt that their interests were under threat from the ideas of Durkheim (actual situation).
C: The group of intellectuals in question polemicized against the ideas of Durkheim.

In fact, here we have an explanation which from a logical perspective seems to be watertight: if we take as our starting point the fact that we can spell out completely the factors in brackets in P1, then C can be totally explained through a combination of P1 and P2. Therefore, the social sciences (and history as well)

11 See for a broader discussion of this topic Hans Dooremalen, Herman de Regt & Maurice Schouten, *Exploring Humans: An Introduction to the Philosophy of the Social Sciences*, Amsterdam 2007.

12 Rein de Wilde, 'Verklaren in de sociale wetenschappen', in: Michiel Korthals (ed.), *Wetenschapsleer. Filosofisch en maatschappelijk perspectief op de natuur- en sociaal-culturele wetenschappen*, Meppel 1989, 94-111. For didactic reasons, I slightly adapt De Wilde's example.

should according to Hempel also have to look for this sort of causal explanation. And just like the natural sciences they should be able to predict certain events. Indeed if we only know that P1 and P2 are true, then we can deduce from them that C will necessarily occur. But, in practice, Hempel observed, sociological and historical studies hardly ever manage to achieve these sorts of watertight explanations, if at all. Clearly the explanatory power of the social sciences is minimal, because human relationships are usually too complex to be reduced to this sort of unambiguous causal link. This means that the sciences concerned do not comply with the norms with which they are supposed to comply.

However, the question is whether this is a correct conclusion. If we take a closer look at the above explanation, it turns out that it is in fact very problematical. For it is of course impossible to test P2 empirically. To do this we would have to have access to what the particular group of critics felt (so that we can confirm that they felt that their interests were under threat). But how could we gain this access? The critics themselves will presumably not confirm their supposed feelings on this matter.[13] They will say: we are not polemicizing against Durkheim's ideas because we feel that our interests are under threat, but simply because his ideas are incorrect, and we want to bring the truth to light!

Now of course, others may counter that what the critics are in fact doing by polemicizing against Durkheim's ideas does come down to defending their interests which are under threat. Whatever they themselves say about the matter, this is in fact their intention. However, such a statement about intentions does not take us any further. The statement 'Durkheim's critics defended their interests because they had the intention of defending their interests (and had the chance to do this etc.)' in fact tells us as little as 'John is a bachelor, because he is unmarried'. For 'X has the intention of doing A' in this connection simply means 'X will do A whenever X gets the chance'. Such a so-called intentional explanation therefore explains nothing. To be sure, we can interpret 'intention' differently, i.e. as a way of referring to underlying motives or *reasons*. In that sense we could assert for example: 'Durkheim's critics polemicized against him, because they had the intention of defending their interests'. By referring in this way to a motive that is at work, we do in fact give an explanation – but once again one that we cannot prove, because it is just as impossible to gain independent access to intentions as it is to feelings.

13 Moreover, if they did do this, that would not tell us the whole story: for there may in fact still be other factors that (also) led to the dispute. What is more, social scientific explanations lose relevance when they only repeat what those involved in the dispute say. In cases of such repetition our knowledge is only increased if other explanatory factors can be excluded.

Nevertheless, since in the social sciences we by definition have to deal with human activities, we also have to deal with intentions in the sense of motivating reasons. And reasons are not causes.[14] A cause offers (provided that it is comprehensive) an adequate explanation for the events that are caused by it. A reason, by contrast, never offers an adequate explanation for the action prompted by it. There is always more at stake, namely the decision of the person in question to allow precisely this particular reason to be determinative in his or her actions. Such a decision can to a large extent be predicted (for example, on the basis of our knowledge of someone's character, circumstances etc.), but can never be completely so. For this reason social-scientific and historical explanations can never attain to the rigour of a deductive-nomological causal explanation.[15] This has nothing to do with all sorts of practical complications, which is what Hempel thought, but rather with problems of a structural nature, or more precisely: with the fact that the relevant disciplines are concerned with the actions of *human beings*.

It is even questionable in this regard whether it is in fact possible to formulate universal laws such as P1 in the social sciences and humanities. If we take P1 as our example again, then we can never completely exclude the possibility that people who feel that their interests are threatened by certain ideas and who have the chance to polemicize against them, will not in fact do so. In such a case, if we cannot find a further explanation, we can only say that they are behaving irrationally. Of course, we could exclude this possibility by adding to the standing conditions that the law is only applicable in situations where people are behaving rationally. So, in addition to people having to have the chance to polemicize, and not to be sick etc. they must also act rationally – only in a situation such as this is the law concerned universally valid.

We in fact do something truly remarkable, however, if we want to save the universal character of the law in this way: we bring ourselves into play. That is to say, in our search for explanations, we unavoidably introduce our own judgements about what is or is not rational behaviour. Not only do we analyse as well as we can the situation of the protagonists and the way in which they themselves judge their own situation; but we also evaluate, from our own point of view, the quality of the reasons that determine their actions (or make them refrain from

14 Cf. Scott Sehon, *Teleological Realism: Mind, Agency, and Explanation*, Cambridge, MA 2005.

15 Only when neuro-scientists manage to convince us that human activities can and should be explained by fully reducing them to specific activities in the brain will this situation change. In the background, however, there are complex philosophical questions at work concerning free will, determinism, the relationship between mind and body etc., that we shall not address here.

acting).[16] In other words, we provide an explanation not so much by pointing to strictly causal relationships, but by interpretation. That is to say: we arrange all the factors that play a role in our description in such a way that all the pieces of the puzzle fit together as closely as possible. In so doing, we exchange the explicative method where we ourselves remain as it were onlookers and outsiders (such as in the research on heated copper wires), for a *verstehende*, or hermeneutical approach, where to an extent we play an interpretative role in the reconstruction of events.

Also when we abandon attempts to formulate universal social scientific laws, this move to a hermeneutical approach seems in practice to be unavoidable. In the study of human behaviour and human activity we simply cannot avoid comparing different factors that play a significant role in social processes with one another. Such evaluations can in the final analysis not be justified in a way that is empirically demonstrable, because human intentions, motives and reasons can never be completely reduced to external (measurable) factors, and therefore their relationship to more objective 'causes' cannot be established beyond doubt.

Let us take as an illustration again the Durkheim example we have already discussed: how did it happen that Durkheim won the argument, so that sociology gained an important place in the French academic world? In response to that question, we cannot give a causal explanation; intentions, reasons and motives play too big a role here.[17] We can only try to formulate as sound an interpretation as possible of the course of events. One of these interpretations revolves around the fact that the interest group of which Durkheim was a leading representative managed to achieve a degree of power. Although Durkheim perhaps had the weakest arguments, from a political perspective this group was stronger or smarter than the circle of literary critics, and they drew the short straw. Another interpretation has it, though, that in terms of content Durkheim had the best arguments in the controversy and as a result more and more people became convinced that his vision was right. And so over time he was able to win the argument. There is material that can be adduced for both interpretations, but the decision over which one we choose cannot in the final analysis be separated from personal judgements, that in turn are rooted in our view of life.[18] So, (neo-)Marxist-

16 It is especially cultural anthropologists who can tell us the extent to which what people consider to be rational or irrational is culturally determined, and therefore how risky this enterprise is.

17 In this regard the Durkheim example is still relatively clear and well set-out. In the case of more complex social or historical events (such as for example the advent of the Second World War) it is that much more impossible to fit causes, reasons, and individual and collective motives into a deductive-nomological explicative model.

18 Karl Mannheim demonstrated that this was 'even' the case with Max Weber, who like no other wanted to erect a defence against all relativism in sociology by (though

inspired researchers, accustomed as they are to think in terms of a struggle between social groups and classes with conflicting interests, will be more inclined towards the first interpretation, whilst by contrast liberal progressive thinkers will tend to plump for the second.

We conclude therefore that the social sciences cannot meet the demands of the deductive-nomological explicative model. In other words: in the area of the social sciences no explanations are possible that are both empirically verifiable and have the form of a logically conclusive argument. It is rather the case that an explanation in the social sciences „... has the character of a 'story' in which human activity is interpreted and placed in an understandable context".[19] And here two or more stories about the same social-scientific facts can exist side by side, each with their own strong and weak points. In the neo-positivistic tradition this situation is considered intolerable. For this reason, those in that tradition used to hold fast to the idea that there was only one scientific method, and tried as far as possible to fit the social and historical sciences into this pattern. But given that they had to admit that it was a failure, they concluded from this that the social sciences are not in fact sciences at all. Indeed, since the so-called social sciences do not manage to attain to objective knowledge, in the final analysis they leave us in the realm of subjective opinions rather than hard facts.

Here, however, we stumble across a particular misunderstanding in (neo-)positivistic thought that persists with remarkable stubbornness in the modern standard picture of science. This is the idea that we have to do with an either/or situation: either the results of scientific research can be proven objectively, or we fall into complete subjectivism. But it is just this all-or-nothing thinking that from a structural point of view has been challenged in the newer philosophy of science. Since the arrival of Kuhn we have been aware that from an epistemological perspective results of scientific research always have an inter-subjective character. In other words: they are judged on how plausible they are by the community of specialized scientific researchers. This does not just happen at random, but always on the basis of comparison: it is a question of whether the plausibility of a certain theory (or explanation) is greater than that of the available alternatives. Even though this cannot be verified objectively, we are still not necessarily then at the mercy of pure chance. Indeed by meaningful debate on the basis of sound arguments a high degree of inter-subjective agreement is possible.

recognizing that the starting point for researchers was to an extent determined by their set of values) seeking an objective research method. See V. Meja and N. Stehr (eds.), *Der Streit um die Wissenssoziologie* 1, Frankfurt 1982, 325-370.

19 De Wilde, 'Verklaren', 110.

It is interesting to note that such an inter-subjective judgement of the results of research does not only turn out to be appropriate in the social sciences, but also equally so in the natural sciences. Here, too, the deductive-nomological explicative model has turned out to be unsustainable[20] from a theoretical perspective and often not to work in practice, so that strictly speaking there are no objective results here. Here as well, it turns out that the formation of theories is of a strongly hermeneutical nature: people create, as it were, a story, that places the facts which require explanation in a meaningful relationship, so that we can understand how they relate to a larger set of theories which already exist. In addition, here as well several stories can stand side by side for a longer time, each with their own strengths and weaknesses (just as in physics with the wave theory and the particle theory concerning the nature of light). Here too meaningful discussion about their strengths and weaknesses is possible.

Wolfhart Pannenberg has pointed out as early as in the seventies of the last century that in the light of all this, we cannot talk of a fundamental methodological opposition between the natural sciences and the humanities. In both cases, the data from individual observations gain significance because they are embedded within a context that gives them meaning. The desire for a deductive explanation arises above all according to Pannenberg whenever an original *verstehen* is undermined by certain developments.[21] In response to this, the original framework which gives meaning must then be clarified or adapted, and this can be done with the help of a natural scientific explanation which shows how certain individual facts fall within a general law. But what is fundamental also in the natural sciences is the search for a coherence of meaning which is as over-arching as possible. This means that 'the study of meaning cannot in itself be the private reserve of the human sciences and cannot justify a theoretical opposition between the human and natural sciences'.[22]

20 For an overview of the many criticisms that have been put forward against it, see Van Holten, *Explanation*, 46-56; Salmon, 'Four Decades'; and Leon Horsten, Igor Douven & Erik Weber, *Wetenschapsfilosofie*, Assen 2007, 127-130. According to Van Holten the situation in the natural sciences is at the moment such that in spite of their exalted status as the explicative disciplines par excellence „... there is simply no agreement on the question of how 'explanation' should be understood: a normative model is lacking" (*ibid.*, Ph D thesis edition, 271).

21 Wolfhart Pannenberg, *Theology and the Philosophy of Science*, Philadelphia (Pa.) 1976, 153-5 (with reference to O.F. Bollnow).

22 Pannenberg, *Theology*, 134 (with 'human sciences' corresponding to what we refer to as the humanities). As we shall see in chapter 6, the same is true according to Pannenberg in theology, which can therefore be conceived of as a scientific discipline. See also the summary in Gunter Wenz, *Wolfhart Pannenbergs Systematische Theologie*, Göttingen 2003, 25f. (including a useful passage on the place that the category of 'das Ganze' assumes in this regard in Pannenberg's writings).

Should we now draw the ironic conclusion from all this that the neo-positivists were right to assert that there is only one scientific method, but, contrary to what they thought, that this of all things is the hermeneutical method? Surely this conclusion is not totally incorrect. Dilthey has in some sense killed his thousands: his brilliant insight into the legitimacy of the hermeneutical method turned out to have a much wider influence than he could ever have imagined. As we saw in the previous chapter, one of the characteristics of the newer philosophy of science is in fact the shift from a formalist (directed towards the formal justification of disparate chunks of knowledge) to a contextualist (directed towards an embeddedness which gives meaning in an all-encompassing context) position[23]. This shift came about as a result of the reflection on the nature and methods of the *natural* sciences. To puzzle in a Kuhnian paradigm or to perform reconstruction work in a Lakatosian research programme are also typically hermeneutical activities, i.e. activities aimed at conferring meaning.

Nevertheless, it would be incorrect to assert on the basis of this that there is no significant difference at all between the natural sciences on the one hand and the social and historical-literary sciences on the other. For in the latter disciplines we have to deal with a two-stage hermeneutical task[24]. We interpret social and historical processes just as we do physical, chemical, biological etc. processes, viz. with the help of theories whose aim is to show relationships between entities. All relevant factors are weighed up as far as possible and placed within the greater whole. The cultural scientist, however, at the same time has to deal with the fact that the objects of his or her study already do this themselves! In contrast to cells or atoms, people already give meaning to their actions and to the reality that surrounds them. In the formation of scientific theories, justice will have to be done to this. For this reason, the position of a participant, whereby the researcher empathizes as far as possible with the object of his study, seems to be unavoidable in the cultural sciences. In this connection there is much debate about how this relates to the position of the observer. Some (such as the social scientist Peter Winch) believe that we must limit ourselves to the imagined position of the participant; others, such as Max Weber assert instead that this position needs to be supplemented by that of the observer[25]. In terms of our example: as Durkheim and his critics do not themselves interpret their differences in terms of a conflict of interests, then

23 As we have seen (section 2.4), this ‘holism’ is an important difference between the older and newer philosophy of science; compare for example Clayton, *Explanation from Physics to Theology*, 18-58.

24 Anthony Giddens, *New Rules of Sociological Method,*, London 1977, 79 (‘The conceptual schemes of the social sciences therefore express a *double hermeneutic* ...’).

25 De Wilde, ‘Verklaren in de sociale wetenschappen’, 106. In the study of religion a similar debate is raging over an ‘emic’ versus an ‘epic’ approach.

the researcher who does do that imposes, incorrectly according to Winch, her own theoretical construction on the object of her research, whilst according to Weber this can in fact provide us with a suitable analysis.

It is clear that in each case this extra complication in the cultural sciences leads to the formation of more schools of thought and a greater variation in research methods than in the natural sciences. In other words: the proportion of the community that ascribes to a given scientific paradigm is much smaller here and there seem to be more competing paradigms that keep the research community as a whole divided. This also has to do with the fact that, to use the terminology coined by Rickert, the individualizing tendency is greater here than the generalizing one. We see this most clearly in the case of history, that deals with the explanation of all sorts of events which are in theory unique[26]. With regard to the natural sciences, Willem Drees speaks of a certain core of 'consolidated knowledge'[27], and it has to be acknowledged that this has less currency in the cultural sciences – although we must not underestimate the body of more or less incontestable facts in, for example, history.

Even so, it is not so much a matter of structural differences but rather of gradual ones, which arise within the overall search for relationships that confer meaning. If there is one characteristic that all sciences have in common, then it is this search for meaningfulness.[28] In actual practice, of course, a range of methodical skills and criteria can be discerned which differ considerably, depending on the area of the discipline. But all of these have the same goal. In this connection it seems plausible to suggest that methodological and theoretical (taken together: paradigmatic) pluralism is gradually increasing throughout the various sciences as the area in which meaning is conferred becomes more comprehensive and thus more complex. We see, in the words of Mary Hesse, a 'linear continuum' between sciences which are more empirically oriented and those which are more hermeneutically so, and any watertight distinction that people want to make between them seems artificial.[29] In the following section, we want to develop this idea further and support it by means of another philosophical contribution to the debate.

26 The historian 'must allow for the exceptional or particular, for only then can the past be seen as the history of one-off individuals, who acted and thought in such and such a way'. H. W. von der Dunk, 'Geschiedenis en wetenschap', in: Cliteur et al. (ed.), *Overtuigend bewijs*, 48.

27 W. B. Drees, *Creation: From Nothing Until Now*, London 2002, 66.

28 Compare Pannenberg, *Theology and Philosophy of Science*, 154, 223f.

29 M. B. Hesse, *Revolutions and Reconstructions in the Philosophy of Science*, Brighton 1980, 225; Van Holten, *Explanation*, 69. Compare also Clayton, *Explanation*, 87f.

3.4 The lasting significance of the theory of modal aspects

In the 1930s, the Amsterdam professors of philosophy H. Dooyeweerd and D. H. Th. Vollenhoven provided the basis for a new Christian-philosophical school, that would become known by the name of 'the philosophy of the cosmonomic idea'. Their work provoked a great national and international response, and even today disciples of them are still developing their ideas in often specialist discussions.[30] One of the most intriguing and most enduring contributions by Dooyeweerd and Vollenhoven to philosophical discussion concerns the foundational debate about how the different academic disciplines relate to one another. We come across these contribution in the so-called 'theory of modal spheres', also known as the theory of modalities or (in Dooyeweerd's work) the theory of modal aspects. 'Modal' comes from the Latin *modus* in the sense of 'manner' or 'way in which'. The theory of modalities deals with the various ways in which we can look at things, ways that are considered to correspond to ways in which things exist. To put it another way: it deals with the various modes of being or aspects or functions that can be distinguished in objects, people, events, relationships etc.[31]

Dooyeweerd and Vollenhoven have dared to suggest that the number of these functions or aspects is not only finite, but is also in principle knowable. What is more, according to them these modalities correspond to the various *sciences* that we have. These are all significant claims that are difficult to prove, but which Dooyeweerd and Vollenhoven made plausible in an elegant way. The metaphor that they use here is that of a ladder, where every 'higher' aspect presupposes all 'lower' ones, without being reduced to them. The order in which they place the various aspects is therefore not in the least random. The characteristic of a lower aspect can be understood without needing the higher aspects, but the reverse is not also true. In this way, Dooyeweerd and Vollenhoven distinguish exactly fifteen modalities for all things (people, events etc.)

The idea fifteen aspects can be discerned in reality and not fourteen or sixteen is of course remarkable. However, those who ascribe to the theory of modalities have always said: whoever believes that there are more or less than fifteen, let them prove it. Their method of identifying the various aspects is therefore strongly

30 For example, in their journal *Philosophia Reformata*. Fundamental to the whole movement is Dooyeweerd's four-part *A New Critique of Theoretical Thought*, Amsterdam 1953-1958.

31 René van Woudenberg argued that the word which Vollenhoven used, 'functions' (*functies*),is preferable to '(modal) aspects' because the meaning of the latter concept is not easy to grasp. 'Functions' on the other hand are said to refer to the intrinsic and relational characteristics of things. See his „'Aspects' and 'Functions' of Individual Things", *Philosophia Reformata 68* (2003), 1-13.

phenomenological (which is interesting given the strong neo-Kantian tone of the rest of their philosophy). The way in which things appear in our experience is determinative for what we ought to say about them from a philosophical perspective. To give the reader an idea of where this takes them, we shall now summarize the fifteen aspects, and in each case indicate how they identify the main characteristic – the so-called core „meaning kernel" or semantic nucleus – of each particular aspect, and also which branches of science correspond to them. Because the names that Dooyeweerd and his group used for the various aspects and semantic nuclei vary quite a bit, I make a particular choice in each case.

Name	*Semantic nucleus*	*Science*
1. Quantitative	number	mathematics, arithmetic
2. Spatial	space	geometry
3. Kinematic	motion	kinematics
4. Physical	causality/energy	physics
5. Biotic	life	biology
6. Sensory/psychic	feeling/consciousness	psychology
7. Logical/analytic	distinction	logic
8. Historical	cultural development	the study of history
9. Linguistic	symbolical signification	linguistics, letters
10. Social	interaction	sociology, cultural anthropology etc.
11. Economic	profit	economics
12. Aesthetic	harmony	aesthetics
13. Justitial	retribution	jurisprudence
14. Ethical	goodness/love	ethics
15. Fiduciary	belief/ultimate meaning	theology/philosophy

Dooyeweerd (limiting ourselves to him from now on) asserted that for example the quantitative aspect is more fundamental than the spatial, because numbers do not require space whilst space does require number. Every space for example consists of three dimensions, has a certain size, possibly a certain number of corners etc. In a similar way, motion presupposes space, but we can think of space without motion, whilst, using modern day terminology, the psychic or sensory domain is *emergent* on the biotic: consciousness (or the human 'mind') is perhaps not fully reducible to neurological processes, although it cannot be dissociated from them. So, for each of the fifteen modal aspects, Dooyeweerd provides one or more arguments for why a particular modality in his opinion belongs in a given place on the ladder and no other.

Each of those arguments is of course open to being tested and debated, and many of his disciples make choices which are different from those of Dooyeweerd. Is there not for example much to recommend the idea that the aesthetic modality pre-supposes the juridical rather than vice versa (so that the beautiful and the good are close together, in accordance with the classical view)? Or why could the aesthetic and the justitial aspects not have a common origin, without one being more fundamental than the other?[32] Is there not also much to support the idea that we can distinguish a sexual aspect close to the biotic and the sensory ones, with sexuality as the 'meaning kernel' – an aspect that would then be studied within gender studies? And is the fiduciary aspect not something of an odd-man-out?[33] More generally we can ask ourselves whether the time for this sort of system aimed at giving an overarching account of the world has not passed. Should we not admit that we simply cannot have the final word in this regard?

These sorts of discussions, however, do not detract from the underlying basic idea of the theory of modal spheres, viz. that we experience reality from quite a number of different-but-interrelated perspectives which correspond with *aspects* of reality. Dooyeweerd himself always emphasized that his theory is open to criticism, discussion and improvements. This has to be so for the sole reason that new developments in the various branches of science can have consequences for the actual development of the theory. But this is also the case, because the various aspects in no way stand apart from each other, and this means that a good deal of judgement is required to create the right set of categories. To be sure, each of the aspects is characterized by its own 'modal sphere', which means a number of related laws that apply in each particular case (for example physical laws for the physical, logical laws for the analytic aspect etc.). Moreover, each of the aspects is also characterized by primary subjects which are 'qualified' by them (for example, stones by the physical, plants by the biotic, animals by the

32 For this suggestion, see, René van Woudenberg, *Gelovend denken. Inleiding tot een christelijke filosofie*, Kampen 1992, 116. Chapter 3 of this book offers an excellent summary of the theory of modal spheres as a whole (66-123), which I am grateful for being able to take advantage of here.

33 Dooyeweerd emphatically distinguished this aspect from what he called the 'religious roots' of human existence, to avoid presenting the religious sphere as simply one amongst many others. According to Dooyeweerd the fiduciary function deals with the human involvement in the Origin of temporal existence, an involvement that in his opinion cannot be reduced to the psychic or logical functions. See the discussion on (*inter alia*) the fiduciary function between Van Woudenberg and this author in *Radix* 21 (1995), 154-177.

psychic function and human beings by all higher functions). But this does not take away the fact that the fifteen aspects are related to each other in all sorts of ways, and by dint of this relatedness in their diversity still constitute one reality.

In this regard, Dooyeweerd spoke about *analogies* between the various aspects, which on the one hand can have the character of retrocipations to a 'lower' aspect, and on the other hand anticipations of a higher aspect.[34] An intellectual movement for example is analytically qualified but with a retrocipation to the kinematic aspect. And animal noises are psychically qualified, but to the extent that they have a certain meaning anticipate the linguistic aspect. Or are they in some cases perhaps even linguistically qualified with a retrocipation to the psychic aspect? Further discussion about this is possible, just as it is about all sorts of other possible retro- and anticipations. It is only when we come to the arithmetical aspect that we cannot conceive of any retrocipations – so apparently this is the most fundamental aspect. Likewise, we cannot have any anticipations for the fiduciary aspect – so here the final dimension of reality opens up.

Besides, according to the theory of modalities it is not true that every item (or every person, event etc., although I shall not list these alternatives every time) only operates in one aspect alone. On the contrary, all of them operate in every aspect, whether as subject, or as object. In the case of the former, the item in question operates primarily in conformity with the laws that apply in the relevant aspect. For example, a stone is primarily subject to the laws of physics, a plant to those of biology. Of course a plant has a physical substratum so that the laws of physics apply to it as well, just as those of prior aspects do for similar reasons (a plant is for example one plant and not two). In short, a plant has a 'subject function' in all the aspects from the arithmetical up to and including the biotic. However, in all those aspects from the psychic up to and including the fiduciary the plant instead has an object function: the plant can be the object of particular feelings, has a certain juridical or legal owner, is valued from an aesthetic perspective (as beautiful or less beautiful), and can even be an object in which a certain solidarity of faith is expressed, for example when a plant or flowers are given as a present to a sick member of the community, as happens in many churches every week. In similar ways all things operate in all aspects. So, the theory of modalities not only attempts to bring to light the diversity of reality but at the same time also tries to give expression to the inter-relatedness and unity of it.

We shall not look any further at how the theory of modal aspects has been and is being developed,[35] but rather focus on its significance for how the different acade-

34 These retrocipations and anticipations are of a logical and not a chronological nature.

35 See the overview in Roy Clouser, 'A Sketch of Dooyeweerd's Philosophy of Science', in: Jitse van der Meer (ed.), *Facets of Faith and Science,* Lanham 1996, Vol. 2, 81-

mic disciplines relate to each other. And here the first thing to note is that in the list of fifteen branches of science some of course are missing (chemistry for example), whilst the place that others occupy seems strange. Although in section 3.2 for example we listed logic together with mathematics under the formal sciences, in our account of the theory of modal aspects they are separated from each other by all sorts of intermediate disciplines. This is somewhat surprising, but this is not the place to go into how this comes about or to consider whether it can be reasonably defended. The important thing for us is that behind the theory of modal aspects lies the principle that each branch of science deals with a certain aspect of reality. Because everything 'has' every aspect in one way or another, all the sciences therefore also deal with all of reality, but from their own perspective, their own 'modal point of view'. In this sense the sciences abstract from the multi-coloured reality as a whole, by concentrating on that one aspect of it as far as possible.

To an extent, this recognizes the inherent limit which is characteristic of all sciences. As the pre-eminent physicist from the last century Hendrik Casimir once remarked about his discipline, physical theories are 'approximate descriptions of a limited section of physical phenomena – which themselves are only a limited section of human experience'.[36] All sciences are similarly founded upon abstraction, in other words, on the isolation and isolated study of one aspect of our colourful experiential world, whilst the true impact and significance of that one aspect only becomes clear within the overall whole. The value of such abstraction is of course clear, but this does not take away the fact that all sciences are, in this sense at least, limited.[37]

The concentration of the sciences on one and only one aspect of reality also implies, and this is where we have to focus our attention now, that the sciences cannot be reduced to each other in any way, but in fact have their own right to exist and their own significance. There is thus a structural *variety* of sciences and the sciences are in a certain sense completely equivalent to each other. Dooyeweerd tries to gain acceptance for this insight in a number of ways. First of all, as we have already seen, by showing that the higher aspects pre-suppose the lower ones whilst the reverse is not true: movement pre-supposes spatiality but not vice versa, et cetera. So, at every higher level something essentially new is added.

98. We can ask ourselves if the theory of modal aspects is in fact specifically Christian. Of course, it is inspired by the Christian doctrine of creation, but one does not have to be a Christian to be able to accept it. Indeed you do not need Christian spectacles to see how reality forms a multi-coloured unity-in-diversity.

36 Hendrik B.G. Casimir, *Haphazard Reality. Half a Century of Science*, New York 1983, 33.

37 For a discussion on the limits of science, see Del Ratzsch, *Science & Its Limits*, Downers Grover 2000[2], 92ff.

According to Dooyeweerd, however, there is yet another way in which the irreducibility of the various aspects can be demonstrated. This is by showing that problems which cannot be resolved arise as soon as we try to reduce one aspect to another. If we do try to do this, Dooyeweerd tells us, we shall not be able to avoid becoming trapped in certain *antinomies*, or internal contradictions. A well-know example here concerns the difference between the psychic and the analytic aspects. In the nineteenth century, various thinkers (including John Stuart Mill), reduced logic to psychology. Laws and principles of logic were to be interpreted as psychic states of affairs, or to be exact, as law-like patterns that arise in human thought. The best way of engaging with these in this approach is then by a process of introspection. In their turn the patterns involved were said to go back to the formation of habits, so that they do not have a universal validity. Edmund Husserl, who was initially a supporter of this 'psychologism', as it was called, later (inspired by the well-known mathematician Gottlob Frege) became one of the fiercest critics of this idea. Husserl managed to demonstrate that psychologism is trapped in internal contradictions. For clearly, as the laws of logic are not universally valid, the same is therefore true for everything that is said in accordance with these laws. But then this must also be true for psychologism, given that it also claims to be in accordance with the laws of logic. Although traces of psychologism could still be found in the work of someone such as Bertrand Russell, currently there is no-one who seriously defends it any more.[38]

Dooyeweerd and Vollenhoven believe that likewise we go wrong when we try to explain away any of the modal aspects. Of course it may be true that we need thinkers of the calibre of Husserl to demonstrate this. For the temptation to reduce the multi-coloured nature of reality is quite understandable. Indeed as we have already seen there are many analogies to be drawn between the various aspects. If we consider them to be straightforward repetitions instead of re-appropriations qualified by another aspect, then there may even be the temptation to consider one aspect as the most fundamental and to reduce all others to this one. And this is how over time all manner of 'isms' have grown up, which do no justice to what we could call the aspectual richness of reality, but rather absolutize one particular aspect.

In the light of the summary of the fifteen aspects above, it would be interesting to look further into how each of them has in fact been absolutized in the past or continues to be so today. We shall not do this here, but in order to give an initial

38 Edmund Husserl, *Logical Investigations* I, London 1970. Husserl changed his mind when Frege wrote a critical review of an earlier 'psychologistic' work of his. See e.g. Antony Flew, 'Psychologism' in: idem (ed.), *A Dictionary of Philosophy*, London 1983, second edition, 292.

insight we shall look at some influential schools of thought which absolutize a particular aspect. We can make a start with Pythagoras, who considered *numbers* to be the truth behind visible reality, and saw them as the building blocks for providing an explanation of the world. As such this is perhaps one of the most understandable reductions, if at least it is true that the arithmetical aspect does not have any retrocipations and in that sense is the most basic. Next we can think of Spinoza, in whose geometrical method *space* is the basic category, so much so that nature and God coextend with it.

If we look at more recent reductions, then we might consider *physicalism* which particularly enjoyed its heyday at the end of the nineteenth and in the first half of the twentieth century, but which today has a kind of renaissance in the philosophy of mind.[39] The idea here is that everything that happens is so rigidly determined by causal laws of nature, that there is no room at all left for something like human freedom (let alone divine intervention). Van Woudenberg tells how it was a shock for the young Philip Kohnstamm when he learnt that his teacher J. D. van der Waals, who later won the Nobel Prize for physics, was accustomed to going to church on Sunday. For how can someone who knows the cast iron laws of nature still believe in God? A physicist who earnestly looked at the facts had, according to Kohnstamm, to be a determinist. In other words: such a person had to believe that reality has to be understood completely in terms of elementary particles and their causally determined movements. 'God, love, freedom and beauty: these are the illusions of the sentimental person, who flees the unrelenting searchlight of physics'.[40]

When physicists later discovered that some fundamental laws of nature (especially in quantum mechanics) are not strictly causal, but rather of a statistical nature, physicalism began to lose its powerful aura – although in large swathes of society it continued to be part of the standard picture of science. Today it is often claimed that physicalism follows from new developments in the neuro-sciences, but this can be debated; although it is often held to be a position inspired by empirical discoveries, on closer inspection arguments and motivations for physicalism are usually of a philsophical nature.[41]

Equally influential today is another form of naturalism (i.e. the belief that all phenomena can and must be explained by the natural sciences), viz. the one that is based on the absolutization of the *biotic* aspect. We might think in this regard for example of writers such as Richard Dawkins and Edward O. Wilson, whose

39 Cf. Jaegwon Kim, *Philosophy of Mind*, Oxford 1996.
40 Van Woudenberg, *Gelovend denken*, 72, who also tells us that later Kohnstamm (when he himself became a believer) came to different conclusions.
41 Cf. Guus Labooy, *Freedom and Dispositions*, Frankfurt a.M. 2000, 52; this book as a whole offers a thoughtful critique of contemporary physicalism in its different (reductive and non-reductive) forms.

publications have received much attention in all sorts of popular scientific articles and programmes in today's media. What is common in their way of thinking is that all sorts of phenomena, including the moral and religious, can be totally explained by the struggle for life that the human species is involved in. Mikael Stenmark is quite right to give a lot of attention to this influential neo-Darwinian variant in his study of contemporary absolutizations of the natural sciences.[42]

From the nineteenth century again comes the absolutization of the *historical* aspect in so-called historicism. The pre-dominant idea here is that all our human knowledge and experience is by definition susceptible to change and is therefore untrustworthy. Ernst Troeltsch developed a genuine world view from this historicism, whilst the sociologist Karl Mannheim, mentioned above, considered the view that all our beliefs, values, norms and ideals are bounded by specific times and places as one of the most important insights of modern times. Because we have gained a sense of historicity, we know that all our values etc. are no more than the temporary product of contingent historical circumstances. Dyed-in-the-wool historicists also applied this idea to the formal sciences and the natural sciences. So, according to Oswald Spengler, the principles of logic should have no value outside European culture whilst according to the philosopher of science, Paul Feyerabend, whom we have already mentioned, there is no reason why we should value knowledge from the Western natural sciences more highly than for example Haitian voodoo practices or Chinese herbal remedies. This reduction of the multi-coloured reality to its historical aspect therefore leads to results that stand in complete opposition to those of naturalistic reductions. The antinomy to which historicism leads is not difficult to find: if all our beliefs are historically determined and culturally relative, then so too is our historicism ....

Next, the reductions of higher aspects such as that of religious belief to the *psychic* function (Feuerbach, Freud) and to the *economic* aspect (Marx) are already somewhat outdated. In the final analysis, the justification for these reductions could not be substantiated, so clinging to them became more and more ideological. As we saw in the previous chapter, it was the ease with which psychoanalysts and Marxists were able to explain everything in terms of their theories that led Popper to his suspicion of ideology and to the sceptical position that turned out to be so fruitful. However, this does not take away the fact that there are still thinkers who explain religion completely in terms of the psychic needs

42 Mikael Stenmark, *Scientism. Science, Ethics and Religion.* In this study, which is particularly strong from an analytical point of view, Stenmark shows that naturalism (or, using his terminology: scientism) does not emerge from the sciences themselves, as most of its supporters assert, but goes back to extra-scientific convictions which are of a philosophical nature. See the bibliography of this book (144-148) for the relevant publications by Dawkins, Wilson and others.

of people, or aesthetic values on the basis of the economic substructure of society.[43]

What are definitely not outdated, though, are the reductions to the *linguistic* and *social* aspects. Since the so-called 'linguistic turn' in the middle of the twentieth century, language has taken centre stage in philosophical enquiry. According to some leading philosophers who have made this turn to language (e.g. Jacques Derrida), we never come into contact with reality as such, but only reality as it is linguistically interpreted and evaluated. So, language does not so much mediate our contact with reality, as was always thought before, but rather blocks it. All higher (and also lower?) aspects are to be found, if we look at things correctly, much more in our linguistic structures than in reality.

Related to this reduction to the linguistic aspect is the so-called social constructionism (or constructivism) of Kenneth Gergen and his followers. According to this view, which itself serves as a new paradigm, our scientific, aesthetic, moral, religious and other sorts of convictions are constructions of meaning which are principally local and context-specific. We always have to bear in mind that our belief claims have their background in our specific cultural milieu. In fact, they never extend beyond this, given the vital insight that the 'truth' and 'significance' of every stock of ideas are in the final analysis to be understood as locally defined, cultural constructions. Thus, we can not distinguish between right and wrong interpretations, for our interpretations simply reflect our own social context (which for that matter, one would think, will also have to be true of our social-constructionist interpretations).[44] We may wonder what consequences this type of reductionism has for morality and ethics.

Finally, there are the well-known reductions to the analytic and aesthetic aspects. We see the former in all sorts of forms of rationalism – the idea, from Descartes to Husserl and the logical positivists, that with the help of universal reason, we can get a hold on all of reality – whilst the latter (aestheticism) arises from the reduction of, for example, religious experiences to the aesthetic realm. Listing to a sermon, for example, can be enjoyed as an aesthetic rather than a religious experience. Is a reduction of other aspects to the fiduciary aspect also possible? This is a question hardly worth asking today, in view of the fanaticism with which for example Islamist terrorists make everything that is beautiful and good

43 Compare for the first example, Robert Segal, 'In Defence of Reductionism', in: R. T. McCutcheon (ed.), *The Insider/Outsider Problem in the Study of Religion*, London 1999, 139-163.

44 See Kenneth R. Gergen, *Realities and Relationships. Soundings in Social Constructionism*, Cambridge 1994; Gergen denies that social constructionism leads to philosophical relativism, because philosophical relativism is also considered as a social construction (108).

in reality subordinate to their faith and to the way in which, as they see it, their faith is treated by others. Faith then draws the moral, juridical and even the biotic dimensions of life up into itself.

It is true that Dooyeweerd and his followers do not provide decisive arguments in all of these cases to bring to light the futility of the reductions and absolutizations. But if we survey the many absolutizations which have been defended in the course of history, the antinomies in which they became entangled from time to time, the way in which they were resolved by new absolutizations of yet other aspects, then it becomes more than likely that we shall find ourselves in a dead end when, in whatever way, we devalue the multi-coloured nature of reality. This conclusion is all the more pressing because all of these absolutizations are fundamentally incompatible. No-one can for example at the same time be both a neo-Darwinist and a historical relativist or social-constructionist, for in each of these cases one will want to reduce the wide range of phenomena to another allegedly fundamental given.[45]

That this sort of reduction and absolutization continues to pop up, however, can be easily explained from the point of view of the theory of modal spheres by the many analogies in the form of retro- and anticipations that exist between the various aspects. It is undeniable that the higher aspects have their substratum in all lower ones. People cannot for example pray – to take a fiduciarily qualified act – without all sorts of biochemical processes in the brain working properly. This fact can make it tempting for example for neuro-scientists, who as a result of their work are often strongly impressed by the extent to which our behaviour is determined by the workings of our brains, to come to the reductionist conclusion: praying is 'therefore' nothing but certain processes in the brain working properly. But this 'therefore' does not sound right. For clearly, the legitimacy of this 'nothing but' reasoning (or 'nothing buttery') is not at all proven. What is more: we simply do not do justice to an activity such as praying as a phenomenon by explaining it merely on the basis of certain functions of the brain.

Of course this is not to say that the reverse has been or can be proven – i. e. that the fiduciary aspect denotes a domain of its own and that prayer can therefore only be interpreted at best in terms of our human commitment to God or to the divine. However, taking everything into account, there is much to commend

45 Of course we can shorten the list of aspects to a much smaller number, by, for example, placing the social aspect straight after the biotic and it straight after the physical and then relating the others to these. Social relationships are then a function of the fight for survival and life itself can be reduced to matter and energy. In cases such as this we could speak of *partial* reductionism. It does not make much difference, for in this example we are simply being physical naturalists: everything that exists is in the final analysis matter and energy.

the view that in scientific research it is crucial to do as much justice as possible to the multi-coloured and multi-layered nature of reality as a whole. The theory of modal aspects, in other words does not offer us a neutral Archimedean point of view, or an indubitable foundation of knowledge, but it does offer us a strong paradigm, or perhaps better still a *model* that can be made very plausible. That model shows that we must not try to reduce the diversity in the sciences to one fundamental science, after which all others should be fashioned. Rather, the methods of the various sciences should cohere with the specific features of the aspect that they study. The fact that there are many methodological similarities, especially between sciences which study aspects that are close to each other (such as physics and biology) should not surprise us given the analogies between the aspects. However, this does not take away the fact that we must continue to respect legitimate differences in methods and results between the various sciences.

To a great extent, those differences are related to the fact that we have to do with what Alister McGrath, taking his lead from Roy Bhaskar, calls a 'layered reality', whereby every layer must be seen as 'real', and open to research with the help of methods that suit the particular nature of that layer.[46] And here the higher up the ladder we go, as it were, the more complex the situation becomes. Every time we go up, there is in fact a new aspect to study or a new layer, which pre-supposes all previous ones. Viewed like this, it should not surprise us that the number of competing research programmes or paradigms is greater at the top of the ladder than at the bottom. Whereas for example statistics has two competing paradigms (the so-called classical school and Bayesian statistics), economics has many more than two and theology possibly has the greatest number and the most persistent ones of all sciences. This is to say nothing about how scientific the disciplines of economics or theology each are, but rather does tell us something about the greater complexity of their various tasks.

Indeed, as the fiduciary aspect opens up the final dimension of reality, then for theology the aim is to provide the meaning of all reality in a comprehensive framework of interpretation. Something similar can also apply to philosophy, to the extent that we might want to consider it as the secular version of theology (in that case, we should perhaps describe the semantic nucleus of the relevant aspect not as 'belief', but as 'ultimate meaning'). In both cases, in my opinion, it is not a matter of expecting 'a straightforward ... empirical approach' from their claims, as Theo Kuipers requires in relation to theology.[47] Things are still less straightfor-

46 A. E. McGrath, *The Science of God. An Introduction to Scientific Theology*, Edinburgh 2004, 139-152, esp. 148. Compare Roy Bhaskar, *A Realist Theory of Science*, London 1997.

47 Theo. A. F. Kuipers, *Structures in Science*, Dordrecht 2001, 248.

ward here than in the natural sciences, and therefore the existential category of personal faith plays a more important role in this connection. That is just how things are, given our limited possibilities for surveying complex reality as a whole and for understanding its meaning.

With these remarks, however, we are in danger of jumping ahead to the later chapters in this book. With respect to the theory of modal aspects of Dooyeweerd and his followers, we conclude that we do not have to subscribe to every detail of it to see that we can be in complete agreement with its tenor. We have good reason to accept that we do not do justice to the multi-coloured reality that we find ourselves in if we explain away some of its constituent aspects by absolutizing others. It comes as no surprise then that we meet this anti-reductionist tendency of the theory of modal aspects *mutatis mutandis* in the work of many other contemporary philosophers, amongst whom we find notable examples such as Ludwig Wittgenstein with his theory of language games. For the standard picture of science this means that it is incorrect to consider the natural sciences as the only true, or purest sciences. In the final section of this chapter, taking our lead from Wittgenstein now, we shall look further at how we must construe the relationship between the various sciences.

3.5 Science as a family-concept

Amongst the different sciences there are important similarities but also important differences. Perhaps it is almost impossible to find a determinative criterion which all sciences and only the sciences fulfil. In that case, a so-called 'essentialist' definition of science is not possible, since there is no uniform 'essence' of science. If we nevertheless want to find some over-arching characteristic, then it is *hermeneutic activity* that above all fits the bill: in all cases it is a matter of giving all sorts of disparate facts a meaningful place by situating them within a wider context. This is what we do with the results of observations in a theory, with theories in a research programme, and with research programmes in a paradigm. The problem is, however, that there are also forms of research that are usually considered to be pseudo-scientific where this happens, such as parapsychology and astrology.

The theory of modal spheres that we discussed above does of course fall short of, as it were, filtering out such pseudo-sciences. So could we perhaps say that *openness to criticism* is a second over-arching characteristic of all sciences, and that this can in fact work as a determinative criterion: parapsychologists and astrologists are too averse to criticism? Unfortunately, this will not work either. For we have already seen that in the natural sciences too a certain 'dogmatism' is often advisable: anyone who rejects their whole theory immediately if the first test gives a negative result is not a good scientist. So, the scientist should be open to criticism, but not too much. The astronomer who rejects Newton's laws because he simply cannot account for the orbit of a planet which has not yet been discovered is too open to criticism. For this reason, we cannot so easily turn 'openness to criticism' into a criterion which can be used to distinguish between science and pseudo-science.

Therefore, it seems to be the case that 'science' is a so-called family-concept. That is to say: just as members of one and the same family usually have certain external characteristics and traits in common with one another, without there being one characteristic that all family members share, so too the sciences, in addition to having unique properties, also have all sorts of methodological characteristics in common. So sciences have certain methodological characteristics in common with some others and share other aspects with still others. The notion of 'family resemblance' in this regard comes from Wittgenstein, who applied it to the meaning of words.[48] According to him, the meanings of words have no fixed, unchanging core either, but do have all sorts of elements that in some contexts are applicable and in others not.[49]

In a similar way we can assume that there is a set of relevant scientific activities and methods which are applicable in certain branches of science and not in others. To this set belong, amongst others: a systematic description of a collection of phenomena which are clearly delineated, the prediction of phenomena, the formulation of theories, the setting up of experiments, the empirical testing of hypotheses, the analysis of sources (whether written or not), the interpretation of texts, the search for inner consistency (= non-contradiction) and for internal as well as external coherence etc. Not all of these characteristics are relevant for all sciences. For example, there is no empirical testing in logic, mathematics and ethics (as we shall see in Chapter 6, there certainly is empirical testing in theology), hardly any interpretation of texts in physics, no setting up of experiments in astronomy, no prediction in the historical sciences and hardly any in the social sciences. So, taking this last example further, we cannot use the simple fact that no certain predictions can be made to argue that theology for example is not a science, unless on the same grounds, we also consider that the historical sciences and the

48 Ludwig Wittgenstein, *Philosophical Investigations*, Oxford 1953, I.67.

49 Wittgenstein, *Philosophical Investigations*, I.65-69, uses as an example the concept of a game. This is used in all sorts of contexts: 'game of skill', 'card game' etc. But here there is no single common essential characteristic that is always present; rather the term has a network of semantic elements and depending on the context, one will be appropriate at one point, another at another point etc.

social sciences as well as a biological theory such as neo-darwinism – where no predictions can be made either – are unscientific.[50]

A few of the other family characteristics that we have mentioned, such as the formulation of theories and the search for consistency and coherence, do in fact apply to all sciences (just like the 'openness to criticism' that we mentioned above), but again they are not truly determinative. We can for example always say that we are searching for coherence, even if we fall well short of it. Conversely, we cannot use the fact of having achieved consistency and coherence as a criterion either, because that would have serious consequences. For example, experiments in quantum mechanics[51] show that light in some respects behaves like particles and in other aspects like waves; but since a system cannot have both series of characteristics together, it remains unclear how they cohere. So the existing 'theory' on the nature of light is incoherent. Few would say, however, that for that reason the theory is not truly scientific.

The only thing that we can therefore do is to establish that disciplines must have *a sufficient number* of these characteristics, in sufficient measure, and have a good mix of them in order to count as scientific. For some disciplines it is absolutely clear that this is the case. For others it is clear that this is not the case (such as for alchemy and astrology); and for yet others there will continue to be disputes for a long time to come (such as for parapsychology and homeopathy). However this may be, there is no reason at all to see the natural sciences as the one and only text book example for all sciences. Rather, there is as rich a variety of scientific disciplines and methods as the reality we live in is variegated and multi-faceted. So the second determinative element that makes up the standard picture of science, i.e. the dominant position that the natural sciences hold within it, is, on closer inspection, untenable.

50 See Richard Swinburne, *The Existence of God*, Oxford 1991, 67; and for the comparison with Darwinism, Michael Banner, *The Justification of Science and the Rationality of Religious Belief*, Oxford 1990, 125-130, 137f.; Van Holten, *Explanation*, 140-142.

51 The interpretation of quantum mechanics is a much-debated issue. See on this e.g. Laura Ruetsche, 'Interpreting Quantum Mechanics', in: Peter Machamer & Michael Silberstein (eds.), *The Blackwell Guide to the Philosophy of Science*, Oxford 2002, 199-226.

4. Perspectivity

4.1 Third characteristic: from fact to theory

In addition to both the characteristics discussed in the previous chapters, the common idea of science that many hold to also includes at least one further element. This is the idea that in essence all scientists set about their work in one and the same way: they observe a number of facts and then develop a theory which can explain these facts in order subsequently to be able to test the theory against the facts. Whenever the theory then turns out to be correct, another new item is added to the totality of our knowledge. It will be clear that this third characteristic of the standard picture of science is closely connected to the two which we have dealt with in chapters 2 and 3. First, it is precisely this supposedly standard method that we have to thank for the fact that science seems to make steady progress. Because it turns out to be possible to observe and subsequently explain ever more facts correctly, ever more blanks in our knowledge are filled in. The first characteristic (the idea of linear progression) is thus based in a certain sense on this supposed 'from fact to theory' method.

A text book example of this method is lightning, a phenomenon that for centuries was interpreted by the most extravagant mythological ideas. But when all the facts could be analysed, the theory of electrical discharge turned out to offer the correct explanation for this phenomenon. At the point at which this became clear, the realm of our knowledge became larger and in this way real progress was achieved. And it is of course not by chance that this example comes from the natural sciences.

The third element of the standard picture is also closely connected to the second in the sense that it is in the natural sciences that we see most clearly how observed facts are explained theoretically, and how the theories that are developed are then tested against the facts. But in the other sciences, so the thought goes, things go on in just the same way. Or at any rate, they *should* go on in the same way – although we should say that the social sciences and certainly the humanities do not unfortunately all manage to do this to the same extent in practice, to put it mildly.

If we look still more closely at how the supposedly standard method in the sciences looks, then we run up against what is called the empirical cycle. The movement 'from observed facts to theory' turns out on closer inspection to be a circular movement that consists of four operations. Following on one after the other they are the observation of facts, the development of theories, the distilling of (testable) hypotheses from these theories, and the testing of the relevant hypotheses against the facts.[1]

Let us take as an example Lakatos' astronomer once more. This astronomer first of all observes that the planet he discovers has an orbit which deviates slightly from what is expected. In truth, we are dealing here with a whole series of observations which support each other time after time: over and over again the movement of the planet in question turns out to deviate from its expected path in the same way and to the same extent. There is thus a certain regular pattern (the deviating path), which calls for a theoretical explanation. But to give such an explanation requires an element of conjecture. In other words: what is determinative here is the creativity of the scientist. We saw that our astronomer certainly did not lack this: he developed the theory that the path of the planet he discovered was influenced by another as yet undiscovered smaller planetoid. From this theory a hypothesis which could be tested was then derived in the form of an actual prediction: if a satellite is put into orbit around the earth, then it should be able to observe the planet which has yet to be discovered, in a location which it should be possible to determine fairly accurately. This hypothesis was then tested against the facts; but it was not confirmed by them, which was sufficient reason for the astronomer then to adapt his theory or (in this case) his hypothesis.

Schematically, we can represent this empirical cycle as follows:

	Theory	
(conjecture)		(prediction)
regular patterns		Hypotheses
(observation)		(testing)
	individual facts	

The various facts and the patterns we find are described with the help of observational terms (such as 'red', 'square' etc.). The processing of these facts in a theory, however, takes place by means of theoretical terms (such as 'gravity', 'electron-spin' etc.).[2] The meaning and significance of the second category can be debated, but not really those of first; for clearly, we can directly observe whether an object

1 In fact, the empirical cycle must be thought of as more of a spiral: it is not the case that we go round and round in circles, but rather that our knowledge gradually increases as in a spiral.

2 For this difference between observational terms and theoretical terms, which in logical positivism is considered extremely important, see R. Carnap, 'Testability and Meaning', *Philosophy of Science 3* (1936), 419-471.

is red or not. The meaning and significance of the theoretical terms, by contrast, can only be gained from testable hypotheses which themselves are distilled from the theory. If when they are tested the hypotheses turn out to be correct then the theoretical terms are vindicated and we have a new discovery; moreover, the scientific nature of the whole undertaking has been demonstrated. For this characteristic *of* science also functions at the same time as a criterion *for* science.

H. J. Adriaanse and his co-authors indeed believe that it is not only the 'testability' but also the 'correctness' of its theories that are amongst the criteria that every discipline which wants to be called scientific must fulfil.[3] If a given discipline wants to be seen as scientific, then the empirical cycle must run successfully and the correctness of its theories must as a result be confirmed. To be sure, it is admitted that a theory is never correct in an absolute sense, but always only in relation to 'the current state of knowledge', and also that no one theory is in agreement with *all* relevant observations, so that we usually have to choose between more or less correct theories. These are of course two rather far-reaching qualifications. But they are not so far-reaching that they would cause Adriaanse and his co-authors to abandon their demand for correctness as such. On the contrary, they stick to this demand, and as a consequence for them for example a discipline such as theology (to look ahead for a moment to what we deal with later in this book) has no scientific status.

However, the question then is whether the current state of knowledge in regard to scientific theory-formation allows for this sort of robust conclusion. From the perspective of the philosophy of science the situation could be rather more dramatic than that one or two minor qualifications of the standard picture would suffice. In any case it is clear from the findings of the newer philosophy of science that also the third element of the standard picture of science does not fully accord with how things work in actual scientific research. Again, the practice of science turns out to be much more complex than was previously thought. The awareness that a good number of insights that were once seen as scientifically correct and in some cases even definitive have had to make room for other views requires us to have a certain humility. According to Geurts, we must 'ascribe just as much of a provisional character to the scientific theories that currently prevail as the theories of old apparently had'.[4]

In the next section, we want to look more closely at a huge difficulty that the belief that the increase in scientific knowledge runs in accordance with the empirical cycle has encountered in the course of time.

3 H. J. Adriaanse, H. A. Krop & L. Leertouwer, *Het verschijnsel theologie. Over de wetenschappelijke status van de theologie*, Amsterdam 1987, 56.

4 J. P. M. Geurts, *Feit en theorie. Kernproblemen in de wetenschapsleer*, Assen 1984^3, 9.

4.2 Observation as theory-laden

It was the American philosopher of science Norwood Russell Hanson (1924-1967; he died young in an accident), who in 1958 introduced a new expression into the philosophy of science debate, when in the first chapter of his book *Patterns of Discovery* he referred to human visual perception as 'theory-laden'. He literally wrote: 'There is a sense, then, in which seeing is a 'theory-laden' undertaking'.[5] Since Hanson coined this expression, it did not disappear anymore from discourse in the field, but it was rather appropriated by an increasing number of authors. Just as everyone since Kuhn started to speak straightaway of paradigms, so too as soon as Hanson introduced the term 'theory-laden' did it become more and more a part of a generally recognized vocabulary. The term quickly became associated not only with visual perception (seeing), but with human perception in general and thence also with perceived facts, so that some people even talked of 'the theory-laden nature of facts'.

In his reflections on the philosophy of science, Hanson wanted to start from the most basic activity that takes place on the scientific shop floor, namely observation. It had occurred to him that differences between competing scientific theories often begin at this most basic level: even the observed phenomena that people claim to interpret are different. Thus, the Danish astronomer Tycho Brahe and his assistant Johannes Kepler saw two different things when they observed a sunrise: the geo-centrist Tycho saw the sun rise over a motionless horizon, whilst the helio-centrist Kepler saw the horizon turning away from the motionless sun.[6] Hanson now argues that this is indeed a question of seeing different things, and not of interpreting the same observations differently. The latter would of course have been a relatively innocent assertion, and hardly a revolutionary one. But what happened was that the different theories of Tycho and Kepler had a fundamental influence on the nature of their observations. As a result, the observations could not function as a sort of independent referee to determine which of them was right. For clearly, the different interpretations could not be tested against the facts which were observed, since there was no unanimity on these facts! The difference in interpretation ran, as it were, right down to the most basic level.

And this, according to Hanson, is how it is with all our observations. 'One does not first soak up an optical pattern and then clamp an interpretation on it,'

5 N. R. Hanson, *Patterns of Discovery*, Cambridge 1972 (originally London 1958), 19; later in the book the variant 'theory-loaded' is used, e.g. 54, 62, 65.

6 Cf. on the relationship between Brahe and Kepler, Peter Barrett, *Science and Theology since Copernicus. The Search for Understanding*, London 2004, 28-29. It should be noted that Tycho was not a geo-centrist in the true sense of the word: his system was a compromise between Ptolemy's and Copernicus'.

he says.[7] So here we are not dealing with a 'first' and a 'then'. In other words, any difference between observing and interpreting is artificial. From the very beginning it is true to say that observing is interpreting; that means, it is the cognitive structuring of an otherwise amorphous patchwork of impressions. For example, we do not hear the notes from an oboe in order then to interpret these as being out of key; no, we hear notes that are out of key – and it is very appropriate that this is how we usually say this.[8]

Hanson tried to show that this is indeed how observation works by using the well-known Gestalt-switch pictures that he came across in the work of Ludwig Wittgenstein. The most famous of these is perhaps still the picture of the duck, which can also be observed as a hare or rabbit (see above, page 54). We spontaneously do either one thing or the other in this case. There is not a sort of objective observation beyond both interpretations, which we can then deal with theoretically in two different ways. Therefore, we cannot say that it is less correct to see a duck in the picture, or adduce arguments for the claim that we should have to see a duck and not a rabbit. For clearly, there is no underlying substratum against which we could test such views and arguments. How is it then that one person sees a duck in the first instance and another sees a rabbit? This has to do with what we *expect* to see first, for example on the basis of what we are familiar with, or on the basis of what information we have been given in advance. If someone says for example, 'Do you see that duck there?' then we shall probably indeed see a duck and not a rabbit – and the reverse is also true. So our observation from the very first moment is theory-laden. That is to say, we are 'directed' by certain regulative principles which we have in our minds, or to put it in technical terms, by our conceptual framework.

Of course this assertion does not imply a boundless relativism. For clearly, in the picture in question we can see either a duck or a rabbit, and even a hare, but, in all fairness, not a rhinoceros or an airplane. Moreover it is unusual in itself that the picture is so ambiguous. Most pictures are not ambiguous, or, at least, not in such an obvious way. This is not because the Gestalt-switch pictures form a rare exception to the rule that we can only observe things in one way, but because for most other pictures we have so to speak only learnt one particular way of observing them at our mother's knee. We could for example just as well see the letter U as a horseshoe, but this does not normally happen because we have been trained to see a character here. Again, it is less obvious to see for example a pen in the letter U.

So it is not the case that our conceptual framework or the theory that we have internalized completely determines what we see. Rather, what we have here is a

7 Hanson, *Patterns of Discovery*, 9.
8 *Patterns of Discovery*, 17.

subtle interaction between on the one hand the reality that we observe and on the other the categories by means of which we observe it, and both of these influence our perception.[9] 'If we want to express this accurately then we should say that observation takes place in a pattern of expectations (...); there is no question of any priority between this conceptual framework and the observed data'.[10] Numerous examples can be given to show that in this way all our observations (and not only that of Gestalt-switch images) are theory-laden indeed. The former Utrecht philosopher of science Geurts, whom we mentioned above, produced a whole number of these examples in a study which has unfortunately all too quickly been forgotten. We shall mention some of them here and supplement them with some others. They all illustrate the same basic idea, which is one we can all too easily either lose sight of or misunderstand. For these reasons it demands our careful attention.

First of all Geurts refers to an experiment by the well-known cultural anthropologist Margaret Mead. Mead once took a native from South-East Asia without much preparation to Singapore and took him round this highly developed city. When she asked him after a while what he had seen, the man replied that he had seen nothing. After being pressed, however, he admitted that this was not completely true: he had seen someone going by with bananas, and he was surprised at the amount of bananas that person was moving. The reaction of the native here seems strange, but in truth it is not. For his past in the forests of South-East Asia had not given him the conceptual means which were necessary to observe what every inhabitant of Singapore automatically did observe. The only thing that he could vaguely relate to observations which were familiar to him was a lorry in the distance packed with bananas! For the rest he could not apply any structure that would enable him to distinguish adequately between the different phenomena and events. Because he could not place his observations within a certain preexisting set of expectations, they remained a jumble for him.[11]

Secondly, we find a similar situation whenever a blind person receives their sight, whether it be through new medical techniques or by a miracle. We find an example of the latter in the Gospel according to Mark, where, in the eighth chapter, a blind man is healed by Jesus. And here the first reaction of the man is very interesting: 'I see people...walking about like trees', he says. Although the writer

9 Talk of categories in this regard reminds us of Immanuel Kant. Kant in fact 'discovered' that the categories of perception are constitutive for all empirical observation; but he still believed that these categories were of a universal nature. Not until the newer philosophy of science did the insight evolve that the nature and influence of the categories of perception are to a large extent culturally and situationally determined.

10 Geurts, *Feit en theorie*, 52f.

11 *Ibid.*, 62.

of the gospel did not of course know of the theory-laden nature of observation, this is precisely the sort of reaction that we might expect on the basis of it. The man who has been cured of blindness can still not at first clearly distinguish between all the impressions that reach his retina. For this reason in his eyes people look like more amorphous organisms such as trees. Only later does his conceptual framework become sharper and he sees people as people.[12]

Thirdly, following on from this, we can also think of the uniform way in which people who belong to another culture were represented in artistic depictions from the past. It is often the case that these representations are not only not very true to nature, but they are furthermore almost completely identical to each other. It seems for example as if one Chinese person looks exactly like another. We can hardly blame this, however, on a lack of attention to detail on the part of the artist in question. For whenever we come across large numbers of people from another culture than our own for the first time, we, too, have the impression that they all look like each other. And it can take an unusually long time before we discover that Chinese people, as far as external appearance goes, are no less different from one another than are, for example, white Europeans. It takes us so long to realize this because our conceptual apparatus is not attuned in the first instance to discerning the many differences that exist between different Chinese people. We only slowly become aware of these differences.

A fourth example has to do not with our visual but with our auditory perception. It is only slowly that medical students come to realize what they should listen out for when they want to hear the heart beat of a foetus with a stethoscope. The first time that they try, they often hear nothing because the sounds they are listening out for are very feint, whilst they have to contend with all sorts of other noises both inside and outside the uterus. After a few months, however, the students know what to do as a result of repeated practice, but also thanks to the instruction of people who have taught them how to distinguish correctly the heart beats of the foetus from all the other sounds, even in a noisy clinic. At this point, they do not provide a new interpretation of the old sounds; rather, they hear new sounds, for now they hear heart beats.[13]

12 See Mark 8, 22-26. The fact that this happens fairly quickly (together with the fact that the man knows what trees look like), shows that the story is not about someone who was born blind – for then it would require years of practice – but about someone who doubtless was able to see in his early days. Looked at in this way, the primary reaction of the man provides an argument for the historicity of the miracle described here. For the parallel development of visual perception after a medical operation, see Geurts, *Feit en theorie*, 63.

13 Geurts, *Feit en theorie*, 62.

Another example that illustrates the point well is the story behind the picture that James Bruce made in the eighteenth century of an African rhinoceros. In his travel journal Bruce tells us that he was familiar with a woodcut of Dürer in which a rhinoceros was depicted, but remarks that this picture had little to do with reality. This was surely true. Dürer, who had to rely completely on accounts from third-parties, had depicted the rhino in 1515 as a sort of armoured dragon – so, as a really exotic animal with an amazing rig-out. Now Bruce asserts that he himself produced a representation of a rhino that is completely unbiased and drawn 'from life'. If we look at the depiction by Bruce, however, we see that it has much more in common with Dürer's woodcut than with real rhinos. Bruce's rhino also has, for example, armoured plating that we just do not find in reality. This in no way means that Bruce could not observe or draw well. Rather, it means that what Bruce in fact saw was co-determined by what he expected to see as a result of his knowledge of Dürer's engraving.[14]

In scientific research too, observations are in part determined and structured in a similar way by existing expectations and theoretical convictions. To put it another way: we unconsciously fit our observations into a pre-existing framework, within which we can give them an unambiguous place. But in doing so we make unconscious choices. We pay attention to some things but not others, or rather: we *see* some things but not others, whilst those other things or aspects that we overlook are no less real (or perhaps, as in the case of Bruce, they are in fact more real). In this way, some critics of the evolutionary paradigm have for a number of years pointed to the fact that clear indications in the direction of a 'design' which lies at the root of our reality have been overlooked.[15] According to them, because we look at reality from within an evolutionary theoretical framework, we hardly see these indications. But anyone who manages to free themselves from it to some extent may suddenly see them as 'stubborn' data, that are difficult to fit into the existing framework.

An interesting question is then of course, how it happens that we often become attached to a given theoretical framework. This without doubt has not only to do with psychological factors, but also with the fact that this framework has often turned out to be *useful* to us. For clearly, by means of it we could solve certain problems that we had come up against – problems that had perhaps turned out to be unsolvable, when we had looked at reality from within another theoretical framework. A nice example of this fact can be found in the different

14 James Bruce, *Travels to Discover the Sources of the Nile*, V, Edinburgh 1790.

15 For a discussion on this, see among many other publications e.g. Robert B. Stewart (ed.), *Intelligent Design. William A. Dembski & Michael Ruse in Dialogue*, Minneapolis 2007.

ways in which colours are differentiated in different cultures. Intuitively, we are inclined to say that we (if we are not colour-blind) can hardly make a mistake in differentiating between colours. Everyone can for example see that such-and-such a vase is green and not blue, can't they? But that's precisely the question. For from cultural-anthropological research it turns out that Navajo-Indians have no separate words for green and blue. They only have one word to denote the colour of both green objects and blue ones. The consequence of this is that they cannot so easily see the difference between green and blue.[16]

So, who sees reality better, who sees it in all its rich colour: the Navajo Indian or, say, the European? We cannot answer this question just like that. For a colour is a *concept*; in other words, it is a sort of mini-theory in our head which we use to structure our sensations. And clearly this can happen in a variety of ways, without our being able to say which one in fact correctly describes reality. In principle, we can carry on endlessly thinking up a different name for every single shade of colour. So we have to make choices and in so doing leave parts of reality underexposed. From this it is also clear why in this regard it does not help us to appeal to the *facts*. For what counts as a fact is co-determined by the concepts that we use! So, a European will for example see it as a fact that grass is, generally speaking, green. But for the Navajo-Indian, things are somewhat different: grass clearly belongs to the category of things that are coloured *bleen*, and it seems strange and difficult to him to deny that fact. (We are using the word 'bleen' – a concatenation of 'blue-green' – here as the equivalent of the Navajo word for all shades of green and blue). Of course the European and the Navajo can discuss this matter with each other in order to see why they differ in their account of the facts. But then it will still not be clear which of them is closest to the 'real' facts.

All in all we have now seen enough examples of what is meant by the theory-ladenness of human perception. What is *not* meant by this term is that everyone who observes sees only what he or she wants to see. For surely there is a 'true' reality which strongly limits the possible observations that we can reasonably have. However, what we observe is not 100% determined by the reality outside ourselves, but also, apart from by our own physical capabilities, by our convictions and expectations – which in turn are again often determined by the language that we have at our disposal. So we cannot simply reject Hanson's thesis by associating it with full-blown non-realism or skepticism, both of which are contradicted again and again

16 See Paul Henle, 'Language, Thought and Culture', in: Henle (ed.), *Language, Thought and Culture*, Ann Arbor 1965, 7. A similar theory concerning Eskimos, who were said (by the cultural anthropologist Frans Boaz in 1911) to have lots of words for different shades of white, has become obsolete by now; see Steven Pinker, *The Language Instinct*, New York 1994.

by our everyday practices.[17] The idea that a reality exists outside ourselves, and that it makes its influence felt, is not in doubt.

So we should not overestimate the significance of the theory-laden nature of our observations. But nor by the same token should we underestimate it either, as if we only mean something trivial by it. This happens, for example, when we see in it nothing more than the well-known fact that people, including scientists, are often prejudiced and are therefore inclined to fit their observations to their theories. Unfortunately this does happen, but here we are dealing with a kind of (self-)deception that we rightly reject. Most scientists are well aware of the fact that they run this risk and so build in sufficient checks and balances to protect themselves from it (for example by describing precisely the circumstances in which they made their observations, so that others can check their findings). With the thesis of the theory-laden nature of observation something more radical is at stake than the prejudices that we have – prejudices which usually can be eliminated if we are self-critical enough. The fact is that our observation is always theory-laden and in this sense *cannot* be neutral or 'unprejudiced'. The easiest way of explaining this is by using a metaphor that we come across already in the work of the Reformer John Calvin, namely that we always apprehend reality through what he calls *a pair of spectacles*. And these spectacles are never completely neutral, but always tinted in one way or another.[18]

If this is the case with our everyday observation, how much more is it so for scientific observations. For generally speaking these are interlaced with theoretical concepts. To give a simple example, it is completely normal for a scientist who is referring to a given observation to make a statement such as 'an electric current runs through this copper wire'. But we can only make and agree with such statements if we consider all sorts of standard theories regarding electric current to be correct. So this description is not theory-neutral. In the description 'an electric current runs through this copper wire' a certain amount of theory formation is already implied, without which we could not even have the observa-

17 In my opinion this happens in the survey of the philosophy of science by Curd & Cover (eds.), *Philosophy of Science. The Central Issues*, 221f. Here it is suggested that the thesis of the theory-laden nature of observation consists in the idea that our perception is *exclusively* determined by our a priori 'beliefs', so that we could never perceive anything other than what our theory prescribes. That idea is clearly wrong – but we cannot dismiss Hanson's thesis so easily.

18 See McGrath, *Foundations of Dialogue in Science and Religion*, 124f. McGrath refers especially to the extended 1543 version of the foreword that Calvin wrote for Olivétan's translation of the New Testament (1534). According to Calvin, the Bible is the pair of spectacles that makes us see the world as God's creation. By the way, in one regard the metaphor of spectacles is not completely adequate for our purposes: people can put on and take off spectacles as they please, whilst the move from one way of looking at reality (or 'world view') to another usually demands more time and effort.

tion in question.[19] And if this is the case with such a simple observation, how much more will it be the case with much more complex observations which are what we usually find in contemporary scientific research.

Once again, all this does not mean that we are left with arbitrariness or full-scale relativism in science. For there is agreement about many theories (although this still does not mean that they are therefore sacrosanct), and where this does not exist different theories can often be meaningfully compared with one another on various points. But the idea that there is an unshakeable foundation of objective observations underpinning the process of theory-formation in science is simply wrong. Although pre-war philosophers of science such as Heinrich Scholz still sought *Voraussetzungslosigkeit* (freedom from presuppositions) in science, today this noble ideal is generally held to be untenable.[20] It is impossible to eliminate totally all our presuppositions of a theoretical or other nature (such as those determined by our view of life), and to deduce our theories directly from the facts. Therefore, the 'empirical cycle' that we described above gives too simple a picture of how things work in science, or how they should work. In contrast to what the standard picture of science suggests, scientists do not take as their starting point facts which are independent of theories, but they work just as much 'from theory to fact' as 'from fact to theory'. Theories are, as the Duhem-Quine thesis has already made clear, *underdetermined* by the facts. That is to say, there is no single way from the facts to one and only one theory, but rather there are always several theories that are more or less compatible with the relevant observations.

Here again we come across the great significance of *hermeneutics*. For it is not a matter of whether a certain theory can be directly checked by simply comparing it with the 'facts', but rather of whether it is capable of helping us to *understand* as many relevant data as possible more than its potential competitors, and of providing us with insight into their underlying relationships. Something like this cannot of course be easily tested. It is often a question of taking time and sometimes also of an unavoidably subjective evaluation of whether the anomalies that one theory encounters are more serious and fundamental than those with which another theory is confronted. In any case the deciding factor is the capacity to solve problems that the different theories possess.

And here we have naturally moved from the terminology of Hanson back to the jargon that was introduced by Kuhn. So it should be no surprise that Hanson

19 See Samir Okasha, *Philosophy of Science. A Very Short Introduction*, Oxford 2002, 89. Also a naturalist such as W. B. Drees recognizes that 'scientific theories are not in a strict sense implied by the data', see his *Creation: From Nothing Until Now*, 68.

20 See esp. his unpublished article, 'Der Begriff der voraussetzungslosen Wissenschaft' (1933), and compare Arie L. Molendijk, *Aus dem Dunklen ins Helle*, Amsterdam 1991, 231-261.

was an important precursor of Kuhn's. Kuhn developed Hanson's views about the theory-laden nature of observation further and in a certain sense we can even say that here we come up against the hard core of Kuhn's own philosophy of science. All sorts of innovative ideas in Kuhn's philosophy of science go back in the final analysis or in any case are connected to the discovery of the theory-laden nature of observation. This discovery therefore forms an important background to the change in the philosophy of science that we described in the previous chapters in more general terms. What is more, in the discovery of the theory-laden nature of observation we see reflected, as it were, on a micro-level what we see on a macro-level as the shift from modernity to post-modernity.

In order to gauge the impact of this discovery more accurately, in the next section we shall look at a few of its most notable consequences.

4.3 Exit: the Theory of the Neat Dividing-line

Since the period of the Enlightenment, a two-fold division has arisen in the West between faith and reason, in which each has been limited to structurally different domains. Faith was considered to be based on traditions, whilst science was seen as a matter of getting rid of traditions and thinking for oneself. The motto that Immanuel Kant gave to the Enlightenment was telling in this connection: *sapere aude* – dare to think for yourself. Kant sent out a rallying cry to the European citizens of his time, urging them to free themselves from their lack of self-determination, for which they only had themselves to blame. 'Enlightenment is man's emergence from his self-imposed nonage ['Unmündigkeit']. Nonage is the inability to use one's own understanding without another's guidance. This nonage is self-imposed if its cause lies not in lack of understanding but in indecision and lack of courage to use one's own mind without another's guidance. *Dare to know! (Sapere aude)* "Have the courage to use your own understanding," is therefore the motto of the Enlightenment.'[21] Especially the repeated use of the phrase 'without another's guidance' is telling here. People should not put their faith in what they hear others from the past or present say, they should only trust their own understanding of the facts.

We can see the enduring power of this ideal well into the twentieth century in a statement by Bertrand Russell:

> We owe the triumphs of science to the replacement of authority by perception and logical thought. Every attempt to breathe new life into intellectual matters with authority is a step backwards (…). One of the great benefits of science for those who

21 Immanuel Kant, 'Beantwortung der Frage: Was ist Aufklärung?', *Berlinische Monatschrift*, December 1784; this English translation by Peter Gay, *The Enlightenment. A Comprehensive Anthology,* New York 1973, 384.

understand its spirit is that science puts them in a position to be able to live without the deceptive support of subjective authority.[22]

This quote neatly reflects the two-fold division which has become very popular since the Enlightenment: in science we make progress thanks to our own powers of perception and thought, whereas faith in authority belongs to a completely different domain.

On the one hand, this way of thinking manifested itself in the Enlightenment period in the increasing criticism of all sorts of established traditions (not least amongst which were those of the church and religion), and on the other hand in the attempt to arrive at a purely rational science. We see the first of these for example in Voltaire and the second already in Descartes. But both of these tendencies are closely related to one another. Together they form the two sides of what I would call the *Theory of the Neat Dividing-line* (or TND for short) – a theory of which thinkers such as Kant and Russell were exemplary proponents.[23] On one side of the dividing-line, we find everything that is based on tradition, faith or authority, and on the other everything that is based on the individual's rational thought. And it is considered to be indecent if we mix the two up. Anyone who does this simply has not fully understood what reality is like. Therefore, we must take the utmost care in keeping these two spheres apart and honouring the dividing line as well as we can. To do this, all sorts of pairs of concepts were developed, and the members of each pair are to be understood as mutually exclusive and even diametrically opposed. If we list a number of these pairs of concepts, then it looks something like this:

Theory of the Neat Dividing-line	
Believing upon authority	Thinking for oneself
Engagement	Neutrality/Freedom from value judgments
Tradition	Reason
Revelation	Experience
Subjectivity	Objectivity
Conjectures	Truth
Norms	Facts
View of Life	Science

22 Bertrand Russell, *The Impact of Science on Society*, London 1952, 110f.

23 I use capital letters here to indicate that we are not dealing so much here with a theory as with a meta-theory: a regulative idea, a perspective, or better still a 'mind set' that is developed concretely in all sorts of ways (which usually reinforce each other).

Of course, we have placed dissimilar ideas on top of each other in the left and right hand columns. This is clear from the fact that they are in tension with one another in all sorts of discourses. So for example in philosophy there are discussions about the relationship between reason and experience in the sphere of epistemology, and in theology different positions are taken up when we are dealing with the relationship between revelation and tradition.

It is typical of the climate of the Enlightenment, however, that these ideas (and we could extend these lists without much difficulty) are more or less listed in this way under and opposite one another. In its purest form TND holds that Athens and Jerusalem have nothing to do with each other and that for this reason a water-tight division must be established between those ideas on the left and those on the right. For things get problematic as soon as they are mixed with each other, or to put it more precisely: as soon as the ideas on the right-hand side are contaminated with those on the left. The statements of Kant and Russell that we have just discussed are paradigmatic in this regard: to make progress we have to throw off the yoke of a heteronomous faith-upon-authority, and instead make the radical choice in favour of autonomous self-determination by thinking for oneself.

Let us now, however, take a closer look at some of these pairs of (supposed) oppositions in turn. To begin with, 'thinking' here stands diametrically opposite to 'believing'. But can we really think without believing? Did the numerous philosophies of the Enlightenment that arose from this ideal not mutually contradict themselves so much that the promise of one solid edifice of universally accepted rational truths could not be fulfilled? At the beginning of the 1980s, the American philosophers Alvin Plantinga and Nicholas Wolterstorff coined the term *foundationalism* for the deep-rooted assumption that they encountered far and wide in the history of Western thought, that the human mind can come to a real knowledge of the truth when we start from a solid 'rock-bottom', an indubitable point of departure.[24] Some years earlier Wolterstorff had already identified the goal of foundationalism as 'the formation of a body of theories from which all prejudice, bias and unjustified conjecture have been eliminated'.[25] To put it in our own words: to come to a position from which all forms of *belief* have been filtered out.

Since then the term foundationalism has been appropriated by countless philosophers from all sorts of backgrounds. And it is now generally accepted that the foundationalist world view has turned out to be incorrect and outdated. Surely,

24 See the collection of essays edited by both of them which has quickly become a classic: *Faith and Rationality. Reason and Belief in God*; see especially Plantinga's essay 'Reason and Belief in God', 16-93. Plantinga and Wolterstorff saw foundationalism as most evident during and after the Enlightenment, but they showed that its roots go back much further than this.

25 Wolterstorff, *Reason within the Bounds of Religion*, 24.

there is a difference of opinion as to what should replace it – some people want to stay closer to classical foundationalism than others.[26] But the feeling that we always and necessarily approach reality from a certain prejudice, a certain belief or rather a certain *perspective*, and that we can change this perspective and correct it but never completely free ourselves from it – this feeling is widely shared in contemporary philosophical circles. In brief: the sphere of 'thinking' cannot so easily be isolated from that of believing as people have thought since the Enlightenment.

We have seen already how this works in science in a very practical way: to be able to get going as a scientist and to be able to make a career of it, you have to begin by taking all sorts of theories on trust without being able to verify them yourself. In short you have to make a certain pre-existing perspective your own and place things within it and use it to make sense of your own work. Of course there are usually very good reasons for making this perspective your own, and accepting upon authority precisely *these* things from *these* people – often scientists with a good reputation! But clearly, this can also be the case in the area of religious belief. A Christian for example does not believe all sorts of things at random, but only specific things which seem trustworthy to him – even when he *thinks* about them as impartially as possible – for example because he has taken them from trustworthy people. So, all in all, it is forcing things too much to place thinking on the correct side of the dividing-line, the side of rational science, and belief on the wrong side, the side of irrational prejudice. If we look closely at their relationship, the parallelism between science and belief turns out to be much greater than TND allows for.[27] In both areas, thinking as well as believing play a crucial role.

Something similar is also true for the other concepts that are often placed opposite each other in TND. Far from being a neutral affair, science for example is usually an activity whose exponents are very passionate and engaged. Often, the most passionate scientists are even the best ones. We can hardly call their efforts 'neutral' because they are driven and inspired by certain values and ideals. This already begins with the fact that they believe that their work is worth the effort in

26 See above, section 2.2.

27 For a well-informed elaboration of this parallelism, see Banner, *The Justification of Science and the Rationality of Religious Belief.* Incidentally, for the sphere of belief this parallelism means that it does not rest on proveable foundational knowledge either. If the church confesses her faith to be indubitable, then 'indubitable' must not be understood as a modal term but rather as a classical gerundive: faith can be doubted – that happens everywhere – but according to Christian understanding it is right and proper not to doubt it. Meanwhile, religious believers are usually much more aware than adherents to e.g. scientism that their faith perspective cannot (in the sublunary world) be regarded as verifiable 'objective knowledge'.

one way or another, and that for this reason it is good from time to time to make sacrifices for it. Perhaps this is because they make a good living from it, or simply because they take pleasure in doing it. But it is also often believed that a scientific approach to the problems of our society is in the final analysis the best thing for everyone and that it therefore makes sense to dedicate oneself to it. Obviously this is a belief that from time to time runs into problems. For example, the threat of terror attacks with weapons of mass destruction must provide a considerable challenge to this belief. Nevertheless, this sort of fear can lead us to hold to this view with all the more conviction, just as religious believers often become strengthened and purified by afflictions to their faith.

In addition scientists, generally speaking, also believe whole-heartedly in the specific research in which they are personally engaged and to which they have devoted a great deal of their time and energy. Usually they recognize its significance and attach value to it; people do not carry out their research with blinkers on, but they have their eye on a higher goal – however far that higher goal may be removed from the shop floor itself. Still, we could insist (with amongst others Max Weber) that these sorts of values and ideals stand apart from scientific activity as such, which no doubt should be and is being carried out in the bomb-shelter of a value-free environment. In the former chapter we have seen, however, that the discovery of the theory-laden nature of observation makes this idea very unlikely (if not in fact impossible). However that may be, there are at least as many faithful believers in science as there are outside it.

Of course this does not mean that such faith does not make sense or that it is incorrect, rather just that it is exactly that: faith. We have to acknowledge that the justification of the common conception of science itself can no longer take place by rational means. Rather, anyone who defends the prevailing understanding of science (our 'standard picture') is 'not an anaemic neutral, but a full-blooded believer'.[28] There is nothing wrong with this, but it is as well to recognize it. In the phenomenon of science, as well as in other areas of life, nothing is achieved without some kind of faith commitment.

But isn't there a crucial difference between such scientific faith on the one hand and religious faith or belief on the other? Do not science and philosophy base themselves, in the end, upon reason whilst religious belief and theology are much more a question of *traditions* that have been handed down? Do we not then come to a point here where the watershed cannot be breached? This has in fact quite often been assumed. A text-book example of this view can be found in the

28 Koningsveld, *Het verschijnsel wetenschap*, 86f. (the quotation is on 87). See also for example Padgett, *Science and the Study of God*, 72, who shows that 'science itself does not present us with a complete worldview'.

writings of late philosopher of religion H.G. Hubbeling (but many other examples could be added):

> The difference between philosophy and theology lies in the fact that theology refers to revelation, whereas philosophy acknowledges only the authority of reason. Theology is the critical reflection on one's own religion, which is presented as a logically ordered system. In the foundation of this system it is permitted to refer to revelation and/or ecclesiastical or other traditional authorities. Philosophy tries to give a critical evaluation of the various religious convictions on rational grounds.[29]

This view, however plausible as it may sound, has turned out to be incorrect on two counts. On the one hand it does no justice to the contextual nature of human rationality, and on the other hand neither does it do justice to the incontestable significance of traditions in *all* areas of life. We could point to a whole raft of literature to support both these points – in particular the concept of 'rationality' has recently lost much of its image of universality as a result of many studies which laid bare its contextual nature.[30] What is or is not rational cannot yet so easily be established, and certainly not in the way that logical positivism sought to do in a formal, context-independent manner. The criteria for rationality turn out to be closely connected to the different contexts within which people operate and the activities that they carry out in these contexts. What for one person in a given situation may be completely rational is not necessarily so for another who has had different experiences. Rationality turns out to have many faces, writes van Huyssteen,[31] and it is simply wrong to ascribe to scientific practice a kind of monopoly on rationality. You do not need to be a relativist to recognize that things are rather more complicated than this.

This becomes even clearer when we approach the matter from the other side, i.e. from the concept of *tradition*. In this area it is particularly the studies published by Alasdair MacIntyre during the last few decades that are widely recognized as ground-breaking. In his research on the history of moral philosophy – the branch of philosophy that deals with morality – MacIntyre came to the conclusion that it is impossible to arrive at one given set of moral norms or values on purely rational grounds. It was an old dream of the Enlightenment to be able to show that anyone

29 H. G. Hubbeling, *Principles of the Philosophy of Religion*, Assen/Maastricht 1987, 4.

30 See J. Wentzel van Huyssteen, *The Shaping of Rationality*, Cambridge 1999, and Mikael Stenmark, *Rationality in Science, Religion, and Everyday Life*, Notre Dame 1995 (to mention only these two examples; in both studies the reader will find references to further literature).

31 Van Huyssteen, *Shaping of Rationality*, 2. It is interesting that Van Huyssteen especially relates rationality to the notion of human *responsibility*. Thus conceived, it is not only of importance in the natural sciences, but just as much in religion and theology.

who thinks purely rationally will discover for themselves what the universal norms for good and evil are. Using all sorts of historical examples, MacIntyre showed that this dream was illusory. The Enlightenment project failed in practice time and again, because no agreement could be reached on what these universally valid rational norms are. MacIntyre managed to gain acceptance for the view that the Enlightenment project *had to* fail because the whole idea of a 'universal reason' is based on a myth. His analysis was the more disturbing, because it brought with it a critical diagnosis of where we stand in contemporary society and how far we have lost our way in a number of crucial respects.

From his historical research MacIntyre concluded that the completely neutral and rational moral point of view that the Enlightenment turned out not to exist. Apparently, there is no 'view from nowhere' (Thomas Nagel). We look, think and judge all the time from a given perspective, moral and otherwise. We can of course change our perspective, but we cannot escape from the perspectivity of our thought. And our perspectives, as MacIntyre then argued, are to a great extent determined by the *traditions* within which we operate.[32] In fact, his book forms a persuasive argument of there not being rationality that is not the rationality of some tradition. Traditions have a profound significance for our human thought in general and in our formation of moral judgments in particular. What would be better than acting as if we could transcend the influence of traditions and leave them behind us, would be to make ourselves aware of their precise role and formative value. Enlightenment thought itself ushered in a new tradition of its own, too, rather than the promised victory over any enslavement to tradition. To use one of MacIntyre's own typically long sentences:

> [T]he history of attempts to construct a morality for tradition-free individuals (…) has in its outcome (…) been a history of continuously unresolved disputes, so that there emerges no uncontested and incontestable account of what tradition-independent morality consists in and consequently no neutral set of criteria by means of which the claims of rival and contending traditions could be adjudicated.[33]

It is therefore the community-shaping traditions which both form and pass on our different ideas about rationality and moral norms (such as 'justice'), and there is no getting away from this. As a result, we can only solve our moral problems satisfactorily within the framework of the tradition and the community shaped by it, to which we belong. For this reason MacIntyre makes a plea for a reassessment

32 We come across this emphasis as early as MacIntyre's first, strongly diagnostic study *After Virtue. A Study in Moral Theory*, London 1981; it was developed further in his second larger publication, *Whose Justice? Which Rationality?*, Notre Dame 1988, as well as in his *Three Rival Versions of Moral Inquiry*, Notre Dame 1990.

33 MacIntyre, *Whose Justice? Which Rationality?*, 334.

of the concept of tradition, one which has inspired many others since. In doing so, MacIntyre is not looking as it were to turn the clock back, but rather to do justice to our situation as it really is.

But are we not at risk here of sinking into the quicksand of relativism? For if there is no independent point of view from which to judge traditions (such as the Christian-Augustinian, the Aristotelian, the Kantian etc.) does it not follow that all traditions are of equal value? This is in fact the challenge that MacIntyre faces. At this point, however, he treads a path which reminds us of Thomas Kuhn (although interestingly he makes no mention of him)[34]: Sometimes a given tradition can hit a crisis which not only can be explained convincingly from within another tradition, but which also can be resolved by it. So, not all traditions are by definition equally effective and equally suited to new, unexpected circumstances. MacIntyre himself believes that Enlightenment thought has hit such a crisis which cannot be resolved with its own principles, and he is inclined in his later work to see in the Christian tradition the most coherent and serious alternative to it. Although this cannot be established by means of a universally accepted form of rationality, one can definitely adduce good reasons for such a view.

We saw earlier (with reference to Michael Polanyi) that just as in (moral) philosophy, in science too traditions are of great significance. Only by growing into the tradition of a scientific discipline and accepting on trust all sorts of ideas that circulate in it can one become a good scientist oneself and so take the tradition further. Also in this regard it is wrong to draw a clear dividing-line between a traditionalism which is considered irrational on the one hand and a universally valid rationality on the other. Even in the (natural) sciences we cannot escape from the particularity of traditions that we find ourselves in 'by chance', and there are often several traditions competing with one another which we cannot transcend, because these traditions at least *co*-determine what we consider as rational.

Anyone who acknowledges this point could however retreat to yet another demarcation line – and here we have the next pair of opposites connected with TND. Perhaps traditions are okay, but doesn't faith rest on *revelation*? And is it not here that there is a crucial difference with science, which of course only deals with human *experience*? It is remarkable what a bad press the concept 'revelation' has received. Religious believers are sometimes excluded from discussions in advance because they 'make an appeal to revelation'. And so they break the rules of the game, so the reasoning goes, as these state that we can only appeal to arguments which everyone can access. The distinction that is made here is remarkable be-

34 Stenmark, *Rationality*, 381, notes the 'striking resemblance' between MacIntyre's and Kuhn's concepts of rationality, and he as well is surprised that MacIntyre nowhere in his *Whose Justice? Which Rationality?* refers to Kuhn.

cause revelation, as the word suggest, simply points to what has been revealed, or made public, and is thus open to all. Whilst the term revelation in an expression such as 'the appeal to revelation' is usually associated with someone's private convictions which cannot be justified further, or to use a well-known metaphor, with a closed house, the original meaning of the word in fact points to a thoroughly public affair. In this sense revelation has something to do with public matters, as for example in 'public science'.[35] Similarly, we can speak of a 'public theology'.[36]

So in the context of faith and theology the concept of revelation points to what has become clear and generally accessible rather than to what is closed and idiosyncratic. For example, Alister McGrath lists eight generally accessible phenomena which play a role in Christianity in this connection: Christian source texts, ideas (such as those set out in creeds), patterns of worship, communities, institutional structures, images, doctrinal terms, and religious experiences.[37] The list, as McGrath himself admits, can easily be added to with other data, such as for example narrative practices (faith and life stories). We come across all these phenomena in our modern experience. The texts, ideas, communities etc. are empirically observable. Together they form a complex, multifarious reality, and both separately and in their fundamental interrelatedness they call for something like a scientific explanation. The best explanation is the one which is most capable of doing justice to the whole. Who or what is responsible for the existence of this huge and sophisticated agglomeration of phenomena?

Lots of answers are possible for this question. But one answer which has had an excellent record for centuries up to the present day is that all of these phenomena as they figure in our experience can best be explained as going back to the words and deeds of God which have taken place in history. And this is what we mean by the term 'revelation'. So, anyone who 'appeals to revelation', says nothing more and nothing less than that he has found this answer convincing. Of course we can choose to agree with this or not, but strictly speaking it is nothing strange

35 Translator's note: The Dutch word for revelation is 'openbaring' and that for public is 'openbaar'. So in Dutch these words are cognate whilst in English they are not. Clearly, however, the word 'revelation' is cognate with the word 'reveal' and so like its Dutch synonym it has the potential to refer to something made public.

36 In the English-speaking world this term has a wide usage at the moment; amongst others we see it in an *Online Journal of Public Theology*, an *Abraham Kuyper Center for Public Theology* (2002, Princeton), and a large number of publications amongst which one of the most well-known is W. Storrar & A. R. Morton (eds.), *Public Theology for the 21st Century*, Edinburgh 2004. The term can have more than one meaning – for the most part it does not refer so much to a scientific, but rather to a socially engaged form of theology. A combination of a scientifically responsible and socially engaged way of doing theology is aimed at by the *Beyers Naudé Centre for Public Theology* of the University of Stellenbosch (South Africa).

37 Alister E. McGrath, *The Science of God*, 150, 211-213.

or closed-off or sectarian or unscientific. It is also difficult to see why we should exclude it from all sorts of discussions. For clearly, also the events that are believed to be the words and deeds of God occurred in the everyday world of human experience – whether it was the experience of a calling by Israel's prophets, or the post-Paschal experience of Jesus' disciples that Jesus was alive. We are not saying here that these events, as Wolfhart Pannenberg has suggested, *have* to be considered by any unprejudiced observer as being 'revelation from God'.[38] For we definitely need a certain 'blik' (Richard Hare) or a certain *illumination* to become convinced of this interpretation. But this does not take away the fact that the basic events in question are in a certain sense publicly accessible: people can inform themselves about them and then make a judgement about them: it's a matter for (public) discussion.[39]

In short, revelation and experience do not stand apart from one another, but are in fact closely connected with each other. 'Revelation' is a certain perspective within which someone can place an experience or a series of experiences, based on the conviction that this is the best way to make these experiences transparent and to place them in a meaningful relationship with one another and with the world as a whole. It is fair to say that this is not so very different from what people do in science. For there too observed phenomena play an important role, but they can never be apprehended independently, because perspectival moments are always in play. We only have to remember what we saw earlier in this chapter concerning the theory-laden nature of observation, where this perspectival moment turned out to be present at the most basic level of the scientific project.

At other levels in science as well 'revelation' plays an important role. This is most evident in the so-called context of discovery: suddenly you 'see' how a complex network of data can be explained, the penny drops, the pieces of the puzzle fall into place, the light goes on, a meaningless series of data becomes a meaningful whole. And like Archimedes you cry out: eureka! Countless scientists have already witnessed to the fact that at crucial junctures their work is marked by these sorts of *Aha-Erlebnisse* (Aha-experiences). At the same time we have seen that according to

38 See (also for a later refinement of this suggestion) his *Systematic Theology*, 1, Grand Rapids 1991, 249-252; we shall look at Pannenberg's ideas in more detail in Chapter 6.

39 One can of course also have a similar argument, as the reader has perhaps already recognized, for religions other than Christianity. But they require separate treatment (and then the question arises as to which religion can not only explain its own data but also those of the other religions in the most insightful way – the theme of the so-called *theologia religionum*). It is enough to assert here that out of the many competing 'claims of revelation', we cannot deduce that all of them are incorrect, just as we cannot conclude from the many rival scientific explanations for a given phenomenon that none of these explanations is correct.

the newer philosophy of science the context of discovery is not so far removed from the context of justification. Accordingly, when someone tries to *justify* a certain theory in retrospective, one usually does so by showing how that theory succeeds in elucidating the variety of observational data and in representing them as a coherent whole. The theory functions therefore in a certain way as a search light, a revealing perspective, a 'revelation'. Even if someone thinks that what has been established by the process of justification is only 'the cause' of a certain fact, then such a cause and effect relationship turns out on closer inspection to stand in a more comprehensive context that gives meaning to that specific causal relationship.

So it seems that far from being mutually exclusive, revelation and experience pre-suppose and reinforce each other. Claims of revelation by believers are always connected to empirical experiences, and empirical experiences of scientists can never be interpreted or explained without an element of revelation. The opposite of revelation in both cases is in fact then not experience, but rather human *thought*. Through the centuries, the development of the sciences has been held back to a large degree because people already *thought* they knew how things worked with the paths of the planets and with a thousand and one other things, so they did not need to look at them any more. People thought this for example because Aristotle had said so, or because according to Greek thought the paths of the planets had to be completely circular. People were not open to the way in which reality revealed itself, because they thought they already knew the truth about it a priorily. Aristotle thought for example that women had less teeth than men, but he obviously had not taken the trouble (as Russell once wittily remarked) simply to look into his wife's mouth. Likewise, in the area of faith many people already think they know whether or not God exists, and if so what he is like. As a result of this they run the risk of becoming closed off to the way in which God makes himself present in history. In both cases for better or for worse we create our own reality by thinking; but this can of course only be a pseudo-reality. So in this regard as well, we see more of a striking similarity between faith and science than a neat dividing-line.

We have now 'deconstructed' enough forms in which the TND presents itself to make it plausible that the other ones aren't persuasive either. So we shall not discuss these other ones extensively here. What we have seen so far makes it abundantly clear that we cannot simply place faith and science in opposition to each other as, respectively, a subjective matter and an objective one.[40] In science as we have seen things are neither purely objective, nor are they totally subjective; science is an *intersubjective* affair, in which the shared insights of one's colleagues are important. Even though 'hard' reality places limits on the possibilities of theory formation, pure objectivity remains an ideal which always beckons us

40 Cf. Van Huyssteen, *Shaping of Rationality*, 27.

forth but which can never be attained; in the final analysis we never progress beyond sentences of the type 'all physicists are agreed that ...'

By the same token, it is simply inadequate to maintain that a phenomenon such as Christian belief is a purely subjective matter, nothing more than a question of taste. Christian faith is far too much imbued and tied up with public, historical events for that to be the case. Without the reports of the crucifixion and resurrection of Jesus Christ, for example, it is impossible to explain the existence of something such as 'Christian faith' in a meaningful way. These reports in a certain sense form 'hard', objective data – even if they are open to differing interpretations. So, even here the objective and subjective elements – what is 'given' and what is 'apprehended' – are closely connected with one another. And as a result we can speak in a meaningful way about how reliable the reports in question are, the extent to which we should allow them to correct certain prejudices, in what sort of light they should be read and from what sort of overarching perspective we can best do justice to the complex, multi-coloured reality of faith as a whole.[41]

Neither can we express the difference between science and faith by using simple frameworks such as 'truth' on the one side against 'conjectures', 'suppositions', 'poetry', or 'myth' on the other.[42] We have already seen how problematic the concept of truth in science is when we discussed Karl Popper (section 2.2). If it is always possible that a theory can be falsified at some time in the future, then its truth is never definitively certain. Surely there is much 'consolidated knowledge', where falsification is not very probable, but this knowledge is once again surrounded by a raft of suppositions – one that does not by definition get smaller.

Conversely, we do religions no justice when we put their truth claims as it were in inverted commas. The Christian faith at least is not only concerned with questions of

41 See Philip Clayton & Steven Knapp, 'Is Holistic Justification Enough?', in: W. Mark Richardson & W. J. Wildman (eds.), *Religion & Science. History, Method, Dialogue*, New York & London 1996, 166, where the authors assert that it is important 'to make Christian beliefs available to intersubjective assessment (...), and to connect them with the kinds of possible evidence, that will make them genuinely discussable by a broad community of inquiry that comprises believers and nonbelievers alike'. It is always striking that this challenge is taken much more seriously by both sides in the US than in Western Europe. Although, in the UK a couple of years ago an approximately 800-page study appeared concerning the question of what exactly happened early on the first Easter Sunday (N. T. Wright, *The Resurrection of the Son of God: Christian Origins and the Question of God*, London 2003).

42 See for example Willem B. Drees, *Over wijsbegeerte van de godsdienst en haar object: Twee formules*, Leiden 2002, 20. 'Whilst in philosophy and science it is about truth, in religion it is ... about a *possible myth*'; elsewhere in the lecture he talks in this connection of a 'supposition concerning a secret' (12) and 'a poetic grasp of what cannot be answered rationally' (18).

wise and foolish, but also of what in the final analysis reality is like, so with questions of true and false. Faith, said John Calvin, is a form of knowledge. No doubt it is a very *particular* form, but knowledge nonetheless. In faith it is a matter of what is and what is not the case. Therefore, there is no a priori reason to link concepts such as 'knowledge' and 'truth' exclusively to scientific research. For clearly, in addition to scientific knowledge there are also such things as moral insight and a sense of the aesthetic, which each have a knowledge-character of their own. The same applies to the knowledge of faith. This cannot be reduced to the knowledge-producing processes of science either. But surely it is (as all knowledge) directed towards the truth. Not that believers have an exclusive right to this truth. If they are being honest, they will be the first to recognize that their knowledge is partial (using Paul's word). But they consider themselves ex officio to be friends of the truth.[43] And they are so affected by what or whom they have experienced as the ultimate truth, that their knowledge claims go further than mere suppositions or poetry. Rather, faith offers them a comprehensive *perspective*, within which all sorts of knowledge claims can be placed and understood in their meaning and coherence.

A final variation on the theme where we see the TND at work is the one according to which in science we deal with the world of facts and in faith with the world of norms or values.[44] The idea here usually is that we simply have to bow before the facts, but that we can pick-and-choose our norms by preference. Facts tell us what we ought to accept, whereas values are to a large extent determined by personal choice and opinion. The physics teacher at school will not be happy if his pupils consider Boyle's law with an open mind, or learn to be critical of it, or train themselves to form their own independent opinion about it. They must rather learn to see that the law holds true! By contrast if we are dealing with something like the significance of the Bible or other contents of the faith, then it seems we are no longer dealing with actual truth, but at best with a matter of personal conviction (at worst we are immediately regarded as religious fanatics).[45]

43 Actually, *The Friend of Truth* ('De Waarheidsvriend') is the name of a Dutch religious weekly. People sometimes smile when they hear about this name, which is no doubt due to its early 20th century flavour; but what, we may ask, is more fitting for us humans than being a friend of truth?

44 See for example Stephen J. Gould, *Rocks of Ages. Science and Religion in the Fullness of Life*, New York 1999; for further reference to this book, see below, section 7.3.

45 For this example, see Lesslie Newbigin, *The Gospel in a Pluralist Society*, London 1989, 40-42. See for some recent reflections on the private – public divide in connection with science and religion: Michael G. Parker & Thomas M. Schmidt (eds.), *Scientific Explanation and Religious Belief. Science and Religion in Philosophical and Public Discourse*, Tübingen 2005.

However, in this dichotomy as well we overlook two things. First, in our pluralist society we tend to put too much emphasis on the normative character of statements of belief at the cost of their factual connotations. It is true that statements of belief are related to a certain normative perspective that one either shares or one does not, but it is wrong to think that nothing meaningful can be said about the reasonableness (let alone the truth) of such a perspective. In any case, statements of belief, as we have seen, aim to make a claim on publicly valid truth. They implicitly or explicitly aim to assert *facts* (=interpreted states of affairs). We cannot therefore simply reject them by saying 'that's what you believe, but I happen to believe something else'. Such a remark, which is often seen as the end of any discussion, can at the very most serve as the beginning of one. For example, statements of belief regarding the creation of the world, the election of Israel, the resurrection and coming (back) of Christ are, it is true, statements of *belief*; but without exception they are concerned with questions of fact, such as the question of whether or not Jesus experienced a bodily resurrection, et cetera. So, we can speak meaningfully, as is the case whenever we are dealing with questions of fact, about the correct interpretation (in the case of Jesus' resurrection for example with reference to the reports about it in the gospels).

Secondly, the fact-value dichotomy does not account for the idea that in science too 'facts' are *interpreted* states of affairs. In terms that we introduced earlier in this chapter: they are theory-laden. Facts pre-suppose a certain interpretative conceptual framework, which is formed in part by the tradition in which we stand. The furious attempts of the physics teacher to make his pupils recognize the truth of Boyle's law can be described as attempts to initiate them into such a tradition. Surely most of us agree that we are dealing with a meaningful and even a *true* tradition here; but this tradition as well is based on a certain perspective, a certain way of looking at things, in which observations such as those which lie at the root of Boyle's law are in any case interesting and important. We might also dispute this. We could for example say that we should in no way want to look at reality in such an abstract way, because that can easily distort our natural interaction with reality. But within the current education system this is not a real option. So, Boyle's law is taken as a fact, and the interpreted nature of that fact usually remains underexposed. Something similar happens to a greater extent with respect to such a powerful idea as the theory of evolution. Scientists (whether they are religious believers or not) who have an open mind about the idea that an 'intelligent design' might underpin the natural world or certain processes in it are often fiercely attacked by colleagues for taking such a position. But whether they are wrong or not, with such a thought-experiment they at least expose the perspectival character of evolutionism.

In the next section we want to take a closer look at a few of these more comprehensive critical perspectives which could be taken up in relation to contemporary scientific practice. We will not do so in order to take sides with these so that we can hold certain unwelcome aspects or results of science at bay in some sort of cheap move, but rather to bring the perspectivity that is part of all scientific practice into sharper focus. For clearly the 'perspectivity' of our knowledge claims is a topic that is easily undervalued in our current scientistic climate.

4.4 *Extra scientiam nulla salus?* Some critical perspectives

We have seen that no-one can escape from taking a certain *perspective*, a given framework within which we place our experiences. This is the case as much for empirical experiences as for experiences of a moral, aesthetic and religious nature. In all of these cases, there is no 'view from nowhere', no neutral standpoint which every rationally thinking person should take as their starting point. When it comes to moral, aesthetic and religious questions we are usually well aware of this. And that is why we consider them to be 'normative' (meaning by this that they cannot be determined by an appeal to universally accepted and verified data). When it comes to scientific questions, however, we are usually much less aware of how much our views are influenced by the perspective from which we consider them. That is the reason why in this section we want to take a closer look at some perspectives on science that are less widely accepted than those which we encounter in the standard picture. In doing so we hope to develop a certain sensitivity for the fact that things may be seen quite differently than we are used to seeing them. Moreover, it will become even clearer that in science more is involved than simply coming to conclusions by working from the facts to proveable theories. What we discovered on the micro-level as the theory-laden nature of observation turns out to have a counterpart on the macro-level in the various over-arching, moral, political and ideological perspectives from which we – consciously or unconsciously – look at science and participate in it.

We shall deliberately not look at perspectives that are coloured by religion (although these of course do exist), but rather draw attention to three other sorts of visions which are critical of the standard picture of science. First, we shall consider a social-critical perspective such as the one which arose out of the so-called *Frankfurter Schule*. Secondly, we shall look at conceptions of science as they have developed in feminist circles and thirdly we shall come back once again to the 'anarchism' of Paul Feyerabend.

1. In the first half of the 1960s, a fierce argument was raging in Germany between on the one hand critical rationalist thinkers such as Karl Popper and Hans

Albert and on the other neo-marxist oriented theorists such as Theodor Adorno, Herbert Marcuse and Jürgen Habermas – the latter united in the so-called Frankfurt School.[46] This argument had already been around for quite a while, but did not really ignite until Adorno and Popper both read papers in 1961 at a meeting of the *Deutsche Gesellschaft für Soziologie*. It has become known as the positivism dispute (*Positivismusstreit*), a term coined by Adorno which is somewhat misleading because Adorno's opponents were not (logical) positivists but critical rationalists who, as we have already seen in relation to Popper, were engaged in fundamental criticism of logical positivism.[47] Nevertheless Adorno saw – and here he was right – a clear relationship between these movements in the fact that they both saw modern science as bringing salvation to society. Although formally the debaters were only dealing with questions of method in sociology, in reality they were indeed concerned with the much more fundamental question of the relationship between science and society.

The place of science in society ought, according to Popper, to be determined by a clear division of tasks. Society, through politics, ought to formulate which practical problems people would be happy to see resolved, and science then ought to be concerned exclusively with looking for solutions to these problems. In theory, if we look long enough, we can find technological solutions to all practical problems. But science has nothing meaningful to say about exactly which problems we should try to find a technological solution for first. For this is a question that is closely connected to the values and norms to which we hold, whilst science ought just to set about its task in a *value-free* way. Of course scientists will usually have an opinion about this too, but they will have this as responsible citizens in society and not by dint of their status as scientists. Science, being neutral, can offer no direction in this regard, for then it would become irrational and ideological. It ought to limit itself to finding the correct means and not occupy itself with the goals. Popper had complete trust in the idea that in an 'open society' of free, responsible and emancipated citizens the choice of goal would be made completely rationally, so that the gradual progress of society was guaranteed. According to him, our modern world, in spite of all its shortcomings, is the best that humanity has ever produced.[48]

46 See the monumental study of Rolf Wiggershaus, *The Frankfurt School. Its History, Theories, and Political Significance*, Cambridge (MA) 1994.

47 See Th. W. Adorno and others, *The Positivist Dispute in German Sociology*, London 1976.

48 Leezenberg & De Vries, *Wetenschapsfilosofie*, 188. We remember that apart from the philosophy of science Popper also occupied himself intensely with social philosophy. For his ideas concerning the organization of society, see in particular his *The Open Society and Its Enemies*.

Adorno considered all this very naïve. He found it incomprehensible that a society that had produced Auschwitz could be judged so positively. Furthermore, according to him it is completely impossible to have a value-free society, and a science such as that which Popper advocated will always in practice strengthen the status quo of society. When people obstinately strive for neutrality they in fact become extensions of those who wield power and make the choices in the society in question. And exactly that is not neutral! The choice of merely delivering what society considers desirable at a given moment is in fact a choice in favour of the ruling classes. It is the choice that, just to mention one thing, led to the development of the atomic bomb.

Adorno considers all this as far from rational, and sees in his own social-critical theory the appropriate way to bring science to its senses. Science must not shun the choice and must not avoid taking sides (which in any case it cannot avoid doing!) It must dare to put itself radically at the service of the lower classes, of the weak, the hungry, the exploited and oppressed of this world. It is completely rational not to wrap oneself up in some sort of alleged neutrality, but rather to stand in complete solidarity with them. Adorno was certainly not convinced, however, that a free, open society of responsible citizens would automatically make this choice. According to him, Popper's 'open society' in fact presupposes a typically Western perspective, which maintains the fundamental inequalities in the world.

The extent to which Adorno is still right in this discussion becomes abundantly clear if we look at the expenditure worldwide on medical research. Some time ago Canadian researchers reported that every year less than ten percent of the medical research budget is spent on sicknesses that affect ninety percent of the world's population. So, for example about 84 million dollars is spent worldwide on research into malaria, as against 800 million dollars on research into asthma. The background of course is that a lot of medical research is controlled by the pharmaceutical industry. Because Westerners can pay much more for their medicines, those in charge of pharmaceutical companies prefer to develop (to name but one example) a new drug for lowering cholesterol rather than one to fight sleeping sickness. And for research into ebola substantial funds are only made available when terrorists start to threaten the west with attacks that could lead to an ebola epidemic.[49]

49 These figures are mainly borrowed from the Global Forum for Health Research, *10/90 Report on Health Research 2003-2004*, Geneva 2004 ('10/90' in the title is an allusion to the so-called '10/90 gap', i.e. the situation that in the recent past only about 10% of the world's financial resources for health research was being applied to the health problems of developing countries, where 90% of the world's burden of „preventable mortality" occurred. Cf. for the most recent state of affairs: Mary Anne Burke & Stephen A Matlin (eds.), *Monitoring Financial Flows for Health Research 2008*, Geneva 2008. I am grateful to dr. René Fransen for drawing my attention to these Global Forum reports.

From such disconcerting surveys it is clear that the idea of a value-free science is indeed an illusion. People always use scientific research to serve certain interests; and conversely the results of scientific research exert influence on what people think is important and on the normative choices that they make. The technical intervention that is necessary for an induced abortion for example cannot be called neutral, because it has a deep influence on the development of moral convictions about how we lead our lives.

In part because of their strong rootedness in neo-marxist philosophy, the views of Adorno have not been very influential in leading circles in Western-Europe. We do find, however, certain important kernels of his views on science in the work of his associate at that time, Jürgen Habermas, who is still very influential. Just like Adorno, Habermas challenges the positivistic view that scientific research can be value-free. For at the root of science there lay what he calls 'knowledge interests' (*Erkenntnisinteressen*). According to Habermas knowledge is never obtained just for its own sake; there is always a certain interest or concern involved – not necessarily a private interest, but possibly a societal one. Anyhow, it is from the perspective of this concern that selections are made, that it is established which sorts of knowledge are of interest and which are not, and what sort of research methods are appropriate for them. In this way, whole areas of science are driven by a governing interest. So Habermas as well opposes the objectivist pretence of unambiguous empirical facts, which, it is often assumed, merely require recording and theoretical elaboration. The research methods and objects of a science are inextricably linked with social interests. Without such interests knowledge has no significance. Knowledge of the number of grains of sand on beach x for example has no significance, because it is not linked to the fundamental orientation of an area of science. Knowledge interests, whether they are of a technical, practical or emancipatory nature, are also always linked with certain groups in society, which are prejudiced or in fact disadvantaged by them.[50]

In short, we cannot detach the theory of knowledge from social theory. Habermas' views remind us here of Thomas Kuhn and his emphasis on the influence of social and economic factors on the direction in which science develops. The question of which 'paradigm' floats to the top, depends in part on what society at a given moment wants to hold true and what it will therefore spend money on. For this reason in permissive societies, research on genetic determinants for criminality often does not get high marks, and criminological paradigms that work with them do not have (or did not have?) it easy. In any case, we ought to keep an eye out for these

50 Jürgen Habermas, 'Knowledge and Human Interests: A General Perspective', in: id., *Knowledge and Human Interests*, 2nd ed., London 1978, 308-311.

sorts of connection, and, Habermas says, look from a critical perspective at the 'value-ladenness' of science.

2. We come across a similar emphasis on the interests that are served by prevailing science and on the values it consolidates in different forms of *feminist* critique of science. However, 'interests' is in this regard a term which perhaps suggests too much conscious intent. Feminist critics of science more often highlight the *unconscious* prejudices which lurk behind scientific research and its results. These prejudices are often unconscious because they are determined by the social context in which the scientists in question work. For centuries this context has been sexist and androcentric in nature. This of course was closely related to the fact that the large majority of official scientists have always been male. In our narrative accounts of the history of science the extent that there have been important contributions by women to the development of science (and of course there have been!) has often been minimalized. While in the first instance feminist criticism was directed at this misrepresentation, later on it became more radical in uncovering the 'gender-images' that scientists have for centuries used, i.e. the culturally determined ideas about what is 'typically masculine' and what 'typically feminine'.[51] These images led to a serious narrow-mindedness which has left its traces on the results of their work – both in regard to (1) observed facts and (2) theories, as to (3) the styles and (4) goals of research.

A well-known example of the first comes from the bio-sciences, and concerns the way in which a sperm cell and an egg cell fuse in (amongst others, human) fertilization. The role of the sperm cell here was as a rule described as active, and that of the egg cell passive. The egg cell waits patiently, whilst the sperm cell heroically struggles against the current in order then to penetrate the egg cell wall, leaving behind all its rivals. Whilst the sperm moves under its own steam, it is said that the egg cell *is* transported through the Fallopian tubes, to wait there to see whether or not it will be fertilized. Only very recently has attention been given to the phenomenon that the egg cell, as is evident from microscopic research, energetically contributes to its own fertilization, by as it were grasping and guiding in the sperm cell with the help of finger-like bulges. What is remarkable is that the existence of these bulges had already been known about since 1895, but that the theory was only adapted to this observation in the 1980s. What we have here is a text book example of the theory-laden nature of observation. Culturally determined images of what is typically male and female turned out to structure observation and to colour the language with which it was described. Only when these images were challenged in the second half of the twentieth century was

51 The term 'gender' is used in distinction to 'sex', the latter term being related not to socio-culturally determined differences but to biological male/female differences.

there room to look at the process of fertilization from another perspective and to 'see' other aspects of it.[52]

Secondly, gender-images also can play a crucial role when it comes to the problem of *theory*-choice. Since as we have seen theories are always 'under-determined' by observation – that is: observations do not necessarily lead us towards one particular theory – it is possible that certain non-scientific values are decisive in the selection of theories. We find an example of this in anthropology, where the development of tools in the so-called hunter-gatherer societies is usually attributed to the hunting man, who is thereby said to have taken important steps forward for humankind. The archaeological material that was gathered was highly compatible with this theory. It was just as compatible, however, with the alternative theory that the food-gathering women were in fact the first people responsible for the development of (often organic) tools. This must then have happened from the very moment that people stopped living in forests, where food was immediately at hand, and moved to fields, where they had to carry, dig up, and prepare food. Although both theories can equally be supported by the facts, it is not until very recently that the gynocentric one has been taken seriously.[53]

Thirdly, the research *styles* that were employed were also influenced by ideas about what was typically masculine or feminine. In concrete terms: methods that were associated with femininity were less accepted than research styles which were considered more masculine. The latter are characterized amongst other things by a certain distance as opposed to involvement with regard to the object under investigation, and by the tendency to think in terms of hierarchical relationships. For example, within molecular biology in the middle of the last century, scientists worked with a hierarchical model of cell organization. DNA was seen as the big 'boss' in the cell, and all sorts of other parts of the cell were supposed to follow its instructions. This had been found to be the case in uni-cellular organisms and it was deduced from this that it would also be the case with higher organisms. However, Barbara McClintock (1902-1992), who was considered as an outsider by most of her peers, and who by the way was no feminist, ploughed her own furrow and just concentrated on the cells of higher organisms (in this case maize plants), with an open mind towards the material and above all towards the differences it contained. This involvement led her to a much more complex theory of cell organization. How things work in a cell turned out not

52 Biology and Gender Study Group, 'The Importance of Feminist Critiques for Contemporary Cell Biology', in N. Tuana (ed.), *Feminism and Science*, Bloomington 1989, 172-187.

53 See Helen E. Longino, *Science as Social Knowledge. Values and Objectivity in Scientific Inquiry*, Princeton 1990, 106-108.

only to depend on DNA, but on all sorts of interactions both in the cell itself and between the cell and its environment. Cell organization can therefore be better understood using a 'social' model. But it was about thirty years before the correctness of McClintock's approach was recognized, and she received a Nobel Prize for her work in 1983.[54]

Finally, gender-images make themselves felt in the choice of research *goals*. Evelyn Fox Keller has shown how this was the case in a most fundamental way in the period in which classical physics developed. She did this by drawing on a quite remarkable fact: the rise of classical physics more or less coincided with a period of intense witchhunts. Contrary to what is often thought, for 1400 years or so Christianity did not bother about 'witches', and the fear of and hunting of witches only arose at about the beginning of the Renaissance, reaching a climax in the 17th century.[55]

Keller makes it plausible that this is connected with the newly developed mechanistic way of thinking which was out to *control* nature as far as possible with the help of measurable data. This ideal stood in opposition to the aim of hermetic philosophy, the great competitor of mechanistic science in the 17th century, which had as its goal the union of spirit and nature. In both cases, spirit was seen as masculine and nature as feminine. Whereas the hermetics (with their fascination for alchemy) saw in the magical powers of witches a miracle of divine influence on material reality, the mechanicists associated these powers with the devil himself. Just as wild nature had to be subdued through rational means, so too the expressions of wild sexuality between witches and devils had to be controlled by means of strict criteria and measures. Keller's argument is not aimed at effecting some sort of return to hermetic philosophy, but rather at drawing our attention to the way in which even our most basic scientific goals have come about under the influence of gender-images, and are therefore neither obvious nor objective.[56]

In short, androcentric prejudices have influenced the choice of facts, theories, styles and goals of research. As a result, the development of Western science would have turned out differently (although the question is whether it would in the final analysis have led to different results), if the perspective of scientists had

54 See Evelyn Fox Keller, *A Feeling for the Organism. The Life and Work of Barbara McClintock*, San Francisco 1983.

55 See also A. van de Beek, *Ontmaskering. Christelijk geloof en cultuur*, Zoetermeer 2001, 56f., who points out that witchhunts stopped once modern science was firmly in control.

56 Evelyn Fox Keller, *Reflections on Gender and Science*, New Haven 1985; cf. Leezenberg & De Vries, *Wetenschapsfilosofie*, 44-46. For the way in which rationality has apparently been linked with masculinity historically, see also Genevieve Lloyd, *The Man of Reason*, London 1984.

not been coloured by this androcentrism. We may ask, however, how we should weigh this conclusion from the perspective of the philosophy of science. In the above examples is it not simply a question of 'bad science', where scientists unfortunately overlooked things, but where the consequences of this were corrected later? In that case, there would not be much wrong. Mistakes can happen anywhere, especially when people are insufficiently conscious of their own prejudices. But if they are sufficiently self-critical to eliminate these, then the integrity and neutrality of science will soon be recovered. Expressed in terms that we came across earlier: perhaps the mistakes that are made only affect the context of discovery and not the context of justification of science. If the rules of the game are applied correctly, then from a scientific perspective the gender and personality of the researcher turn out to be irrelevant.

Some feminist critics of science believe that this is indeed the case. Others, however, consider the situation more serious and, following on from thinkers such as Kuhn and Richard Rorty, assert that knowledge never simply rests on the objective representation of reality, but is always theory-laden and is co-determined by the perspective that we take.[57] The researcher who has led the way in this field, Sandra Harding, distinguishes in this regard between three currents within the feminist critique of science.[58] To begin with there is 'feminist empiricism', supported by someone like Helen Longino.[59] This current or strand of thought accepts the standard picture of science and considers the narrowing of vision we have pointed to therefore as a matter of incidents which are regrettable, but which can be put right by the self-purifying power of science. In order to do this, however, scientists must be very aware of their social identity and of the values they represent. Only by being aware of the extent to which we are tied to values can we filter out their influence.

The so-called 'standpoint theory', though, argues that this is in fact impossible. According to this second current of thought, to which Harding herself subscribes, the cultural standpoint or perspective from which we work can never be filtered out from scientific research. The scientist is never an abstract subject, but always a historically and culturally located human-being of flesh and blood. This locatedness

57 For Kuhn, see Chapter 2 above, and for Rorty, see especially his important work *Philosophy and the Mirror of Nature*, Princeton 1980.

58 Sandra Harding, *The Science Question in Feminism*, Ithaca 1986. The three currents do not, though, form the total picture. For example, Kathleen Okruhlik, 'Gender and the Biological Sciences', *Canadian Journal of Philosophy* (Suppl. Volume) 20 (1994), 21-42 takes a middle position between the first and the second current, by seeing in 'openness to criticism' a guarantee for objectivity. According to her, objectivity is not guaranteed methodologically but rather socially.

59 See Longino, *Science as Social Knowledge*.

has an unavoidable influence on the nature and content of our knowledge. And thus the question must be asked: Who is it who in fact knows? It makes a difference whether it is a man or a woman, and even more what type of gender-constructions they hold to.[60] Rather than denying the essential importance of this, we should become aware of our own situatedness, so that we may learn how to take the *correct* standpoint. For the various standpoints are not equally true or adequate! For example, a feminist standpoint leads to better science because from there we can overcome the distorting effects of androcentrism.[61]

But is there in fact one and only one feminist or genuinely female standpoint? The third current, that of 'feminist postmodernism', rejects this idea. In reality there are many fragmented standpoints (amongst which feminist ones, as well as perspectives which are determined by class, race etc.). Each of these leads to a unique story, a unique social construction of reality. But we cannot weigh up these constructions against each other in a rational way, for there is no overarching standpoint. Concepts such as 'truth', 'rationality', 'objective knowledge' etc. belong, so it is said following on from the French postmodernists, to a particular discourse to which they to ascribe validity, but separate from which they have no meaning. There is also then no reason to expect our salvation from the great story of science, any more than from other discourses. But if this is the case, can we still consider a feminist perspective to be better than an androcentric one? Here the feminist critique of science seems to shift into relativism and what is sometimes called 'philosophy of science anarchism'.

This does not take away the fact that the feminist critique of science as a whole is currently taken completely seriously within the official philosophy of science.[62] This is of course also the case because it has not stopped at criticism, but with the discipline of women's studies or gender studies it has developed its own alternative forms of scientific practice. At first sight the idea that someone's sex or gender makes a difference to how one practices science seems very un-

60 See Lynn H. Nelson, *Who Knows? From Quine to a Feminist Empiricism*, Philadelphia 1990; Lorraine Code, *What Can She Know*?, Ithaca 1991. A leading idea in this type of research is that male thought is often directed towards domination (classical mechanics for example was a typically male affair), whilst women are more geared towards being receptive and relational, which would lead to a different type of knowledge. It is of course remarkable that in such a way of thinking traditional images of the differences between men and women are implicitly endorsed.

61 In this connection, Harding appropriates Rorty's pragmaticist concept of knowledge, and thinks that feminist analyses are not so much more correct as more *useful* than others, because they make a contribution to the emancipation of women.

62 In this way the overview we have already mentioned of Curd & Cover, *Philosophy of Science* discusses feminist philosophers of science and evaluates their contribution (170-253) – although it is attempted to incorporate them into the standard discourse.

likely. For science is neutral and value-free, isn't it? But perhaps we think so because the standard picture of science is still left intact in the back of our minds. The gender-perspective then functions as one of the perspectives that disabuse us of the idea that this standard picture is self-evident. One does not have to be a relativist to recognize that what passes for consolidated scientific knowledge sometimes has less objective validity than is suggested.

3. At this point, we could bring up against both of the critical perspectives that we have so far discussed that they are of an ideological nature, because they take neo-marxist and feminist ideas respectively as their starting points. This objection presupposes that in science something like unpartiality is necessary and possible – which is precisely what is at issue. But perhaps we may think that the outspoken social and political commitment of both perspectives makes it difficult in each case to come to an open-minded assessment of the standard picture. For this reason we finally turn to someone whose critical perspective did not come out of an existing ideology and who can be suspected just as little of a religious motivation, namely *Paul Feyerabend* (1924-1994). We could call Feyerabend a sort of lapsed member of the scientistic church. It was not some social or political orientation, but precisely his intensive involvement in the philosophy of science which made him take an increasingly critical position with regard to the whole scientific project and its pretentions.

Having started out as a Viennese logical positivist, Feyerabend then came under the influence of the work of Popper, with whom he also studied for a time in London. He developed into a respected philosopher of science, who participated in all sorts of (often technical) discussions in the broad empiricist tradition. However, in the second half of the 1960s, under the influence of thinkers such as Wittgenstein and Kuhn, he turned his back on Popper's ideas and emerged as one of his harshest critics. As we saw in Chapter 2, Feyerabend especially criticized Popper's belief that there is something like a standard scientific method which ought to be followed in order to make progress in science. According to Feyerabend, any method-monomania leads to a narrowing of vision and thence to stagnation. Because people think along the lines of their own method (and perhaps also of their own favourite theory), they are not therefore open to other, unorthodox ways of approaching a problem. For clearly, as a consequence of the theory-laden nature of observation and the impossibility of falsifying a theory definitively, people can immunize their own methodological and theoretical approach against criticism for ever, even if their approach has in fact become barren. In a detailed historical study, in particular in relation to the Galileo-debate, Feyerabend shows that science often gained important successes when scholars moved away from the standard scientific method, even trampling on the rules of their discipline and in fact therefore carrying out their work in an irrational manner.

Therefore at this stage of his thought, Feyerabend argues for a methodological and theoretical *pluralism*. The motto is: let a thousand flowers bloom, for only then can we escape from the enforced order and uniformity which prevent any substantial progress. Over and above this, a theory can be better (for more rigorously) tested to the extent that there are more alternatives at hand. In this context we should understand Feyerabend's famous expression 'anything goes'. This expression does not voice a kind of skeptical resignation or decadence in relation to the philosophy of science, but rather is a well-thought out maxim, based on thorough-going studies in the history of science (just as we saw this in the case of Kuhn). In his major work *Against Method* Feyerabend develops these ideas even more radically by taking them not only in a prescriptive but also in a descriptive way.[63] Science *is* in essence an anarchistic enterprise, which cannot be regulated by means of laws and rules. Famous scientists are in fact clever opportunists, who made the right choice at the right time. There is therefore no reason to put science on a pedestal, and to consider it better than other (just as non-methodological) forms of acquiring knowledge. Accordingly, Feyerabend no longer believes in a single method and sets himself against the whole scientific project, because as he sees things science wrongly claims a monopoly on knowledge, truth and rationality.

In reality, according to Feyerabend, everything in science hinges on power, just as it was previously the case in the church. Once an instrument of liberation from forces of darkness, science itself has now become an oppressive ideology: anyone who does not agree with its assertions, thankfully is no longer killed – we have become too civilized for this, although no thanks to science – but is instead marginalized. Its claims on knowledge and truth are, as is always the case with ideologies, purely at the service of its own interests. For this reason Feyerabend argues for a strict division between science and state, similar to the one between church and state. Further, he looks for (and flirts from time to time with) non-scientific traditions of knowledge acquisition. As such he points for example – having once been cured himself by alternative medicine – to old Chinese healing techniques (which turned out to work miraculously well when the Communists reintroduced them), but also to astrology, to the still unexplained architecture of the pyramid builders in Egypt as well as to the voodoo cult in Haiti and other magical practices.

63 Paul K. Feyerabend, *Against Method. Outline of an Anarchistic Theory of Knowledge*, London 1975. The book refers to an article with the same title by Feyerabend in 1970, which he had wanted to develop together with Imre Lakatos into a study *For and Against Method*. In this co-production Lakatos was going to argue for the existence and correctness of a uniform scientific method. However, the death of Lakatos in 1974 made this impossible. Later the planned title was used for the publication of the correspondence between the two of them (*For and Against Method*, Chicago 1999).

If science does not work according to a standard rational method and if it therefore has no firm ground, then its status in society ought to be the same as that of this sort of practice. In reality, however, science has become 'the most recent, most aggressive, and most dogmatic religious institution' that one can think of.[64] It has even managed to present itself so forcefully that we intuitively consider it to be much more rational than all other competing practices and institutions. But this is purely a result of 'power play'. Objectively there is no reason for this, just as there is no reason why we should compel pupils in education to become acquainted with all sorts of branches of science and their results (results which, to make things ever worse, are usually presented as facts rather than as views or opinions).

But has science not shown that in contrast to many non-scientific practices, it is able to solve all sorts of problems? The moon landings and the many miracles of medical sciences for example provide the ultimate proof of its unparalleled success, don't they? In answer to this sort of question Feyerabend does not relativize these successes, but he asserts that science could not have achieved any of them under its own steam. For clearly it has no method of its own, and as a result has constantly been intertwined with all sorts of non-scientific factors, which are equally responsible for its successes. Medical science for example owes much to various forms of traditional knowledge from primitive peoples about the effects of herbs et cetera. That it tries to depict this traditional knowledge as a sort of unwitting preliminary stage for science is typical of the way in which it tries to gobble up competing traditions.[65]

In this way, Feyerabend attempts to break away from the bewitching suggestion of Western science that it alone is salvific. Feyerabend expressed this suggestion by rephrasing an old statement by the church father Cyprian. Whereas he expressed the religious belief – which was common in his time – that outside the (catholic) church there is no salvation (*extra ecclesiam nulla salus*), the belief in science that characterizes modern society rests on a precise analogy to this: *extra scientiam nulla salus* – outside science there is no salvation (and, we might add, no knowledge).[66] It was against this belief that Feyerabend ranged all his intellectual and rhetorical powers.

Although it is less easy to refute Feyerabend's ideas than would perhaps seem to be the case at first sight, clearly his work does include some paradoxes and weaknesses. So, for example, in his in-depth studies Feyerabend made use of standard historiographical methods which are recognized as scientific in order to show that there is no such thing as a scientific method. Further, as an avowed

64 Feyerabend, *Against Method*, 15.
65 This account is based in particular on the final chapter of *Against Method*.
66 Feyerabend, *Against Method*, 306.

defender of a cultural and epistemological relativism he did not manage to avoid the paradox of any relativism: if everything is relative (a product of interests, power play etc.), then this also applies to relativism itself. Next, the parallels that Feyerabend draws between established science on the one hand and alternative practices such as astrology and magic on the other may work well rhetorically, but are less convincing with respect to content. And when he finally depicts science as a religious institution based on a (pseudo-)religious belief, he makes insufficient allowance for the structurally different natures and goals of scientific research and religious belief.[67]

But all this does not take away the fact that reading Feyerabend is extraordinarily instructive.[68] It should make any supporter of scientism uncomfortable. For Feyerabend makes abundantly clear what I would like to call the *perspectivist* character of science. In other words, he shows that the prevailing scientific fundamental assumptions are less self-evident than is often thought. For they are based on a certain way of looking at reality, a way which is not neutral in the sense of objectively-valid, nor the only possible way, and certainly not the only way that can lead to knowledge. In a manner that brings him close to today's postmodern philosophers he deconstructs the knowledge and truth claims with which science is linked and criticizes their absoluteness. We do not have to agree with the relativistic implications that Feyerabend drew in order to recognize his significance for our contemporary reflection on the phenomenon and the practice of science. More than anyone else before him, Feyerabend laid bare how close the boundaries between science and ideology are, as well as the large extent to which the scientific way of thinking is a theoretical perspective from which we interpret reality rather than an factual given that arises directly from reality.

4.5 But surely there are proofs?

In this final section we want to discuss an obvious objection to the main theme of this chapter. This objection is that there are in fact *proofs*. We can be impressed for a long time by relativizing considerations such as those in the last section – only suddenly to find that we come crashing back to earth when we run up against the concrete results of science. So observation may be theory-laden, theories underde-

67 See below, Chapter 7.

68 See also the positive comments that Clayton, *Explanation*, 45f. dedicates to Feyerabend. Clayton correctly points out that theologians should not be too eager to join Feyerabend, because religion in his eyes is not one jot better (or more rational) than science. Against this Clayton asserts: 'A good case can be made for significant criteria of method and justification in theology' (46) – which we shall see confirmed in Chapter 6.

termined by the facts, science not value-free and its results not objective – there are in fact all sorts of things that science has incontrovertibly *proven* and that no right-thinking person can get round, or so it seems. We just have to think of a basic fact such as the earth going round the sun and not the sun round the earth. And this is of course just one of the many thousands of things which can be confirmed thanks to scientific research. Should philosophers of science not have to pay much more attention to this sort of hard fact, rather than lose themselves in general and often abstract views about theories, methods, paradigms and the relationship between science and society?

Let us, then, test this objection by trying to find out exactly what a proof is. When can we say that a certain process of argumentation has been successfully concluded so that the hypothesis or theory in question has been proven? One thing is clear: a proof always has the form of an *argument* or of a connected series of arguments, which together form the evidence. Any proof therefore consists of one or more arguments. Yet not every argument is a proof. So the question is: when can an argument be considered a proof? George Mavrodes has pointed out that an argument must satisfy four conditions for this to be so: the argument must be valid, it must be sound, it must in a certain sense be cogent, and it must be convincing.[69] Let us take a closer look at these conditions.

The idea that an argument must be *valid* means that the conclusion must follow logically from the premises. The following argument is for example valid:

Premise (1): All cows are green
Premise (2): Daisy is a cow
Conclusion: Daisy is green

This argument is valid, because it is not logically possible to confirm the premises and at the same time to deny the conclusion. For clearly, the conclusion follows from the premises. But we do not recognize this valid argument as a proof, and rightly so, because (if we take the dictionary definitions of the terms used as our starting point) one of the premises is untrue, namely premise (1). In order to qualify for the status of 'proof', an argument has therefore not only to be valid, but its premises must also be true. If both of these conditions are fulfilled, then an argument is *sound.*

Let us now look at a very well-known argument that is not only valid but also sound:

Premise (1): All human beings are mortal
Premise (2): Socrates is a human being
Conclusion: Socrates is mortal

69 George Mavrodes, *Belief in God. A Study in the Epistemology of Religion*, New York 1970, 22-35. The current section is grafted onto Mavrodes' argument, although the reader will find other example syllogisms in his work.

Can we speak here of a proof? That depends. The conclusion certainly follows on logically from the premises, and it also seems that the premises are true. But, imagine that someone does not agree that both premises are true. It is for example possible that Jews and Christians deduce from the fact that according to the relevant Bible stories, Enoch and Elijah went up to heaven, that these Old Testament figures were not mortal. Muslims could think something similar with respect to the prophet Mohammed. In that case, they will think that premise (1) is false. Still, it is of course possible that we do not doubt the truth of the conclusion for a second. But this will be for reasons other than those adduced in the argument.

In other words: the argument given has no evidential value and therefore cannot count as a proof for someone who is not convinced of the truth of the premises. So in order to be able to be considered as a proof an argument must be not only sound, but people must also *recognize* that the argument is sound. Only in that case is an argument *cogent*. As long as people do not recognize that the conclusion follows logically from the premises or that the premises are true, then the argument is not cogent and cannot qualify as a proof.

Before we take a closer look at whether any cogent argument is a proof, let us first make another observation. There is a difference between the way in which it is established whether arguments are valid and sound, and the way in which we establish whether they are cogent. The former happens on the basis of purely formal logic. If it turns out that an argument is indeed valid and sound, then this is the case for everyone. But the latter is dependent on the person for whom an argument is intended. This person alone can tell us whether he or she recognizes that the premises are true (and the deduction valid). How then do we prevent deadlock if the person in question does not recognize the truth of one of the premises? Well, in that case we can try to prove the premises in a way similar as we did with regard to the conclusion. So we could make the premise in question the conclusion of a previous argument. Such an argument for premise (1) could for example look something like this:

Premise (1): All human beings will die at some moment, unless they are taken up to heaven by special divine intervention.
Premise (2): Even those who have been taken up to heaven alive, would have died at some later moment if that had not happened.
Premise (3): To be mortal means to die at some moment provided that there is no special divine intervention
Conclusion: All human beings are mortal.

It is possible that my (Jewish, Christian, or Muslim) conversation partner will be convinced by this argument that, despite his or her previous reservations, all human beings are indeed mortal. In that case the argument in which 'all people are mortal' functioned as a premise in order to show that Socrates too is mortal, will have evidential value for him or her. But it is also possible that my conversation partner will not be convinced by the above argument, for example because he or she does not agree with premise (2). In that case we can only carry on by trying to demonstrate the truth of premise (2). In order to succeed here perhaps I have to demonstrate the truth of one of the premises which underpin premise (2) – and so on until I reach the point at which I can convince my conversation partner.[70]

From this observation we can now draw the intriguing conclusion that proofs are by definition person-relative. A proof is always a proof *for someone*, i.e. for anyone who experiences the relevant argument as cogent. We recognize this fact of course from jurisprudence: the point at which a judge delivers a judgment is always so tense, because it cannot be established beforehand which arguments for or against the person on trial he or she will consider cogent. If someone does not experience an argument as being cogent, then we can only try to demonstrate the correctness of the premises which are in doubt. And here in principle we come up against an endless regression (a so-called *regressus ad infinitum*). In practice this cannot of course happen. At some point we come across premises which can no longer be proven, but which are still considered true. In mathematics – the discipline in which talk of 'proofs' is still most appropriate – we speak about axioms in this connection. These axioms (such as the one that two points can only be linked with one another by one and only one straight line) must simply be accepted if we are to make any progress. In short, not all our views can be proven, not even in such an exact discipline as mathematics. In general it is true for any given argument that not all of its premises can be proven, because the series of premises is simply unending.

Can we now say that an argument which is cogent for someone also has the qualities of a proof for that person? Let us take a closer look at this by means of the following example, where we accept that this argument is not only valid and sound, but also cogent. So not only the logical deduction is correct and the premises are true, but we also recognize that all of this is so.

Premise (1): If Vienna is not in Austria, then something is wrong with my memory.
Premise (2): Nothing is wrong with my memory.
Conclusion: Vienna is in Austria.

Are we dealing here with a proof? No. The point is that we can only recognize the truth of premise (1) from our knowledge of the conclusion. If we had not been convinced of the fact that Vienna is in Austria, then we could not have been

70 Mavrodes, *Belief in God*, 27, talks in this regard of a 'termination rule': 'The series of challenges and replies will end when a proof is obtained whose soundness is know[n] to the participants in the discussion'.

convinced either that something was wrong with our memories if we had thought that something else was the case. In short, in this case we do not deduce the conclusion from the premises, but rather we deduce premise (1) from the conclusion. Only because we know that Vienna is in Austria are we convinced that something is wrong with our memories if according to that memory Vienna was not in Austria.

So a cogent argument must also be *convincing* for the person at whom it is directed. Only then can we talk of a proof. And an argument is convincing if it is cogent for person N, and if N does not deduce the premises underlying it (directly or indirectly) from the conclusion. At this point, it becomes even clearer what it means to say that proofs are person-relative. What is a proof for one person, does not have to be so at all for another. This depends on the prior knowledge that someone has. We can even imagine that what a premise is for one person forms the conclusion of the proof for the other, and vice versa. The burglar will deduce from the fact that he has stolen the securities that the owner no longer has any securities. The owner will deduce from the fact that he no longer has any securities that they have been stolen. In both cases the argumentation leads to a considerable widening of the knowledge base. But the direction of the argumentation is different in each case, and so therefore is the nature of the proofs.

Mavrodes concludes from all this that argumentation is something like working with *tools*. Tools are only useful when we already have things that we can work on with them. Likewise argumentation is only effective when we already know certain relevant things. From this starting point argumentation can then take us further. A proof is like a mallet that we use to knock a tent peg deeper into the ground. If we know nothing relevant, then an argument cannot help us further, just as the hammer cannot help us any further if there is no tent peg. In other words: proofs are not only person-relative, but also (I would add) *instrumental*. A consequence of this is that no-one can acquire all their knowledge simply by means of argumentation. And if therefore any knowledge at all is possible, then there must be other ways of acquiring knowledge than by argumentation.[71] We can therefore never prove everything that we (think we) know.

If we keep in mind this person-relative and instrumental character of proofs, then we get a clear idea of what is going on in *scientific* argumentation. What exactly does a statement such as, 'It has been scientifically proven that the earth goes round the sun and not vice versa' mean? This statement does not mean that what is being claimed is objectively true because it has been verified once and for all in a way that is universally valid. Rather, it means that there are arguments

71 Mavrodes, *Belief in God*, 41f. In this connection Alvin Plantinga and other so-called Reformed epistemologists coined the term *properly basic beliefs* for beliefs that we cannot, but also do not need to, prove.

(which we may assume to have been derived inter alia from empirical research), which have been convincing for a relevant group of scientists. From the knowledge which they already thought was at their disposal, these scientists have been able to make a cognitive leap forwards with the help of a given argumentation.

The notion of 'proof' in science is therefore an *intersubjective* category and not an objective one. It is important to realize this, for it means that sentences which begin with 'science has shown that ...' can never be intended to silence someone – although they have of course constantly been used for this purpose. Rather, such sentences invite us to ask: *How* has 'science' proven this? – a question which sometimes causes some embarrassment to the speaker. Is what counts as proof for the relevant group of researchers also a proof for me, as someone who perhaps has a different starting point when it comes to prior knowledge? If I have much less relevant knowledge at my disposal than the group of researchers in question, then it is obvious that I am inclined to take them at their word (just as many scientists also do amongst themselves). This does not mean that their proof is also my proof, for my starting point remains different, but it does mean that their trustworthy reputation and their endorsement of each other's findings has evidential value for me. However, if I think I have other knowledge at my disposal, perhaps even unproven knowledge, for example by dint of the fact that within the enormously complicated network of premises I have different views on one particular point, then it is less obvious that I will accept the relevant statement just like that.[72]

So we see that even the notion of a scientific 'proof' cannot save the standard picture of science. The notion is definitely not without significance; it is certainly more than a rhetorical ornament intended to add weight to one's own beliefs, which is what postmodern philosophers believe. But it is unavoidably person-relative and instrumental, and so less 'scientific' than is often suggested in positivistically-inclined circles.

We can now round off our reflections on the standard picture with a three-fold conclusion. First of all, the common idea about the gradual 'progress' of science is at best problematic. Secondly, there is no reason to consider the natural sciences as a methodological text-book example for everything that we may call science (science meaning here any disciplinary tradition that satisfies academic standards). And thirdly, scientific research is not based on 'hard facts' which precede all theory formation, but entails notions such as interpretation, contextuality and perspectivity from the very first moment.

72 This sort of views or beliefs which co-determine our theory-choice is sometimes called 'control beliefs'; so for example Wolterstorff, *Reason with the Bounds of Religion*, 63-66, 72-80.

The fact that proofs are person-relative and instrumental in nature is also of importance in the realm of *faith and theology*. For clearly, a view that is adduced time and time again as an argument against faith and theology is that we cannot *prove* important articles of faith such as the existence of God. Depending on how one conceives it, however, this view is either empty or incorrect. It is empty if what is meant is that the existence of God cannot be confirmed in an objective manner which is convincing for everyone. This is of course true, but the same holds for any other assertion you care to mention. Therefore, the claim that we cannot prove the existence of God, conceived in this way, is empty: we do not succeed in saying something special by uttering it. On the other hand it is incorrect if we mean that there are no arguments that convince *some* people or groups of people of its truth, and that may therefore count as proofs (in the sense that we have been developing here) for them.

I remember that when I was a student, a lecturer in the philosophy of religion wrote a syllogism on the board and in front of a room full of skeptical students kept on stubbornly defending the fact that here was a completely convincing proof of the existence of God. The syllogism he wrote down was as follows:

Premise (1):	Jesus of Nazareth said that God exists (and that it is important to reckon with God)
Premise (2):	Jesus of Nazareth was not a liar
Conclusion:	God exists (and it is important to reckon with God)

It seems to me that the lecturer in question was right. The fact that not everyone will be convinced by his argument is not the point. It was enough for him that some people were (and are) convinced by it.

It is therefore wrong to create an opposition between science and faith around the issue of 'proofs', as if science were based on proofs whilst religious belief has no firm foundation. In this regard the Theory of the Neat Dividing-line does not work either. As in scientific research so in religious faith, things such as convictions, suppositions, doubts, pieces of evidence, impressions etc. play a determinative role. Just as religious belief stands or falls depending on whether or not we are personally convinced and won over by the experiences that we have had, so too the choices that are made in science are closely related to the importance that we attach to the arguments for and against a given theory. Dutch physicist Van den Beukel puts this clearly:

> Whether you 'believe' in the Big Bang today or not, or whether you have doubts about it, depends only on whether you consider the arguments 'for' more important than those 'against'. It is a personal, subjective choice therefore. And even if you think that you can decide 'for' or 'against' today, you had better be prepared for the fact that tomorrow new evidence will appear that throws that decision up in the air

> (…) Just like physics, faith rests on people's experiences. The whole Bible and the Christian tradition are full of people who *testify* that God has entered their lives, that He has called them, that they have heard his voice, that they have lived with Him. They have spoken of and testified to that experience, and they have discovered that they were not alone in this experience, that it was shared by others.[73]

In short, the intersubjectivity that we came across as soon as we began to analyse the notion of a 'scientific proof' in some depth is something that we also meet mutatis mutandis in the domain of faith.

With these final reflections we have already moved on from considering the phenomenon of science as such to questions concerning its relationship to faith and theology. In doing so we move on to the rest of this book, for from the next chapter onwards these questions will intensely occupy us. Our aim will be to consider the consequences of what we have discovered about the current state of affairs in the philosophy of science for the nature and status of theology (chapters 5 and 6), and also for the relationship between science and religious faith (chapter 7). By looking in this way at the relationships between science, faith, and theology as the activity which connects these two we come to the second focus of this study.

73 Van den Beukel, *Waarom ik blijf* [= Why I don't leave the church], Baarn 2003, 153f.

5. Theology and the older philosophy of science

5.1 With what shall we compare theology?

Summarizing once more the results of our journey through the history of the philosophy of science that we began in Chapter 2 and continued in Chapters 3 and 4, we can describe these in very broad terms as follows. We met five schools or movements which broadly speaking came one after the other, though sometimes they also continued or continue to run more or less side by side for a longer period of time. In the table below, in each row we give the name by which the movement in question is best known, the criterion for good scientific practice that it proposes, and the main problem which this criterion turned out to face. Of course, what we have here is a significant simplification and schematization of the much more complex reality; so for example in all the movements many more criteria and problems play a role than those we list here. But those we do list at least provide an important overarching theme through the history of the philosophy of science.[1] Therefore, the table is therefore not totally without meaning.

Name	**Criterion**	**Problem**
1. Logical positivism	Principle of verification	Induction
2. Confirmationism	Principle of confirmation	Demarcation
3. Critical rationalism	Principle of falsification	Adaptation
4. Paradigm theory	Puzzle-solving capacity	Relativism
5. Research-programme theory	Discovery of 'novel facts'	Adaptation

The shift from the first to the second, the second to the third movement etc. demonstrated in each case, as we have seen, a certain logic. So, for example, confirmationism arose (amongst other reasons) in order to help deal with the problem of induction that the original logical positivism came up against. The

1 The recent survey of the philosophy of science in Hans Dooremalen, Herman de Regt & Maurice Schouten, *Exploring Humans*, has basically the same classification of schools, with chapters on logical positivism (161-193), critical rationalism (including confirmationism; 195-236), paradigm theory (including Wittgenstein's notion of language games; 237-294) and 'research programmes & methodological anarchy' (Lakatos and Feyerabend; 295-324).

critical rationalism of Popper attempted to find an answer for the problem of demarcation, which turned out to be simply unsolvable from within confirmationism. But Popper's falsificationism too came up against an extremely stubborn anomaly: that of the problem of adaptation. By this we mean that a scientist can in principle carry on endlessly adapting his theory – including elements of it that are wrong – to the results of empirical research (as Duhem and Quine showed). Falsifying experiments can always be overcome by 'ironing out problems', in other words by calling certain auxiliary hypotheses into being so that the core of the theory is not put at risk. And often it is only rational to do this – but the question is when to do it and when not?

On this point Kuhn asserted that we are forced to remain silent here, and that at most we can give a few broad (and unavoidably subjective) guidelines which may bring to light differences between various paradigms in their ability to adequately solve puzzles. This answer in turn brings up the threat of relativism in the philosophy of science (a relativism that was embraced for example by Feyerabend), which was why Lakatos, in response to Kuhn, sought a better alternative. We discovered, however, that Lakatos' attempt ran aground on the same sort of problem that undermined Popper's approach, namely the possibility of endless adaptation – whereby a certain questionable ad hoc adaptation can still turn out after a long time to be correct, thus vindicating the relevant programme.

It is interesting that the order in which these five movements are presented here is broadly speaking their historical order (although Lakatos wrote not much later than Kuhn, he certainly did expressly react to him in his work), but that this does not mean that we find a continually upward trend here. Indeed, in the final analysis, Lakatos' theory of scientific research programmes comes up against the very same problem as that for which Popper was unable to obtain any convincing solution. Just as we discovered that the history of science does not develop in a linear-progressive manner, this is clearly also the case for the history of the philosophy of science. It seems that all in all we have not progressed much further than Kuhn's conclusion: that in the development of the natural sciences we can talk of an increasing ability to solve puzzles (this can be seen for example in the increasing technological possibilities), but we cannot per se talk of getting closer to the truth or of holding within our grasp how reality does in fact work. Sometimes the secrets of and behind reality seem to become hardly any less; and the risk of collective errors remains an ever-present danger.

If we now move on from here to an analysis of whether or not theology is 'scientific', we shall take as our starting point the table above which sets out the five schools of thought in the philosophy of science which are generally considered to be the most important. This strategy has an obvious reason behind it: anyone who wants to pass judgment on whether or not theology is a scientific discipline will have to be clear about what he or she understands by the term. And this is exactly the question that is at stake within each of the various movements in the philosophy of science, and which in each case has generated different answers. So, for any of these movements we ask the question: imagining that we take as our starting point the concept of science advocated by this school of thought, what then given this definition of science are the chances and possibilities for a scientific theology?

On this point we shall intentionally take a different line from that which Adriaanse and co. took in their study of the scientific status of theology. They too begin their study with a sketch of the history of the concept of science, and they also round it off with a discussion of the newer philosophy of science ushered in by Thomas Kuhn.[2] However, they then concentrate their research on what they call 'the prevailing science paradigm', a little later described as 'the modern Western science paradigm' (34).[3] So, they take as their starting point the idea that in the modern Western world there is one prevailing view as to what science consists of.

We are of course then curious to know which 'science paradigm' they are talking about here. In this regard, they make two remarkable choices. First of all they limit themselves to "that part of the prevailing science paradigm which in many regards forms its central core, the concept of 'theory'" (54). And then in establishing what should be understood by the term 'theory', they take their lead from a basically Popperian methodological handbook for the behavioural sciences by the influential Dutch psychologist A. D. de Groot (1914-2006). The criteria that they use to judge whether or not theology is scientific result then from these two choices.

Both steps seem to me to be illogical and in that sense also open to criticism. First of all by concentrating on the concept of theory, they ignore one of the most envigorating insights that has found general acceptance within the newer philosophy of science, namely that a theory can never be seen as a separate unit, isolated from the greater whole of which it is part. Along with Quine, we can think of this greater whole as a network or web of belief, with Kuhn as a paradigm, or with Lakatos as a research programme – it is essential that a theory never stands on its own, but is always a part of a more comprehensive matrix which to an extent determines its nature, content and pre-suppositions. Secondly, the choice of taking the methodology of De Groot as a guiding principle means that a proposal that precedes the Kuhnian shift in the philosophy of science has been consciously chosen as a reference point. For however authoritative De Groot's methodology has been,

2 H. J. Adriaanse, H. A. Krop, L. Leertouwer, *Het verschijnsel theologie*, 13-33. Numbers in brackets in the main text in this section refer to this study.

3 A science paradigm ('wetenschapsparadigma') is a paradigm (more or less in the Kuhnian sense) that defines what good science consists of.

and however many insights of lasting importance we find in it, his work dates from the early sixties of the last century.[4] Since the rise of the newer philosophy of science we can in no way see it any longer as an example of 'the prevailing scientific paradigm' (54). Add to this that it is asserted that theology should follow the path of the general theory of science at least from some distance (34), and the choice of taking a starting point from 1961 seems all the more strange.[5]

To be sure, later in the book the authors attempt to justify their choice of the science paradigm that they consider to prevail. They rightly consider this necessary for the success of their project and do it by comparing the chosen paradigm with five others. However, these alternatives are not only somewhat weak and under-developed (as the authors themselves acknowledge they are not in fact so much dealing here with comprehensive paradigms as with ideas which only differ slightly from one another), but without exception they also date from the time before the turn to the newer philosophy of science. They refer to thinkers like Edmund Husserl, Max Scheler and others. And when they do consider more recent conversation partners, such as Wolfhart Pannenberg, they do not give an account of the way in which they have appropriated the newer philosophy of science. The confrontation with Kuhn and others, which should have been included here (and which for example was also certainly possible via Pannenberg), is not in the least forthcoming. As a result, they cling to the idea of a universal, non-paradigm-bound concept of rationality, without carefully weighing up or even discussing the reason why this idea was brought into discredit in the newer philosophy of science.[6]

If such an approach could not really be justified in 1987 then now at the beginning of the 21st century the newer philosophy of science seems only to have got stronger, and so it strikes us as that much more unsatisfactory and inadequate. This is why we shall take a different approach in this book – not by trying to follow the

4 A. D. de Groot, *Methodology. Foundations of Inference and Research in the Behavioral Sciences*, The Hague/Paris 1969 (the Dutch original appeared in 1961). H. M. Kuitert seems to go even further when in his *Filosofie van de theologie*, Leiden 1988 he bases the criteria which theology should satisfy on the work of Heinrich Scholz from 1931, which he sees as a benchmark for 'the prevailing concept of science' (27). Even if along with A. L. Molendijk, *Aus dem Dunklen ins Helle*, 324, we place Scholz closer to Popper than to the Vienna circle, this is of course still quite remarkable. However, the way in which Kuitert develops these criteria (in particular Scholz's 'verifiability postulate') turns out to be less positivistic than Scholz had in mind, and leads him to a quite different conclusion about the scientific nature of theology (83ff.).

5 De Groot himself then deals with the later developments in his so-called forum theory of scientific truth, in which he emphatically rejects the idea of an 'objective' scientific truth. See A. D. de Groot & H. Visser, *Het forumwaarmerk van wetenschap*, Amsterdam 2003, VIII.

6 Adriaanse & co., *Het verschijnsel theologie*, 97-118 (published in 1987).

latest fashion in the philosophy of science (which may turn out to be a hype), but by following the developments indeed 'from a distance', taking our starting point in the major shift from the older, prescriptive stage in the theory of science to the newer philosophy of science. As we have said, with each of the five conceptions of science (a term which perhaps is preferable to 'science paradigms', because not all of them are incommensurable) we shall ask the question: if science must indeed be defined as it is on the basis of this conception or view of things, what consequences does this have for theology? Can it be considered as a scientific activity, or is that impossible? And given that it is not out of the question by definition, how should theology then go about its task from a methodological perspective in order to throw light on its (possible) scientific nature? It may be clear that in order to find this out we shall have to pay attention to the answers given to the above questions from the conceptions of science prevailing in the newer philosophy of science. It is, therefore, especially with these that we shall compare theology. And it is only to show the difference in outcomes that in the first instance in this chapter we shall compare theology with the standards of the older philosophy of science. Before we do that, however, in the next section we have to answer a few preliminary questions.

5.2 The relevance of theology as an academic discipline

Prior to our study of the consequences of the earlier and more recent philosophy of science for theology, there are still three questions that we have to discuss in order to put the relevance of our exercise in the correct perspective. First of all, we shall ask in this section what in fact theology is. Secondly, we shall consider why theology should really need to be scientific. Can theologians not do their work well without harbouring this pretension? And thirdly we shall consider whether theology should just let the modern philosophy of science determine its status, or whether it also perhaps has something to say on the matter itself. Why should theologians not be able to make their own contribution from within their discipline to the philosophy of science debate?

First of all, it is important to make clear the manner in which we shall conceive of the concept of theology in what follows. Here too there are of course different possibilities which can all count as 'live options'. For example, theology is often defined as – broadly speaking – the study of the views and behaviour of believers, or adherents to religions. In this sense the term is more or less synonymous with the study of religion, or 'religious studies' as it is nowadays called. The discipline can then again be practised diachronously (history of religion) or synchronously (systematic or comparative study of religion, and next to this also psychology of religion, sociology of religion etc.). In most of these cases the question as to the nature of theological research can be reduced to one which considers the nature of similar research as this takes shape for example in the study of history, psychology or sociology. For clearly, in this case the theologian hardly does anything else than applying the methods of these other disciplines to a somewhat different object (which however does not necessarily call for an essentially different approach), namely the phenomenon of religion. At most we can have a discussion about the status of the comparative study of religion, although this is not very controversial either since it operates like a cultural science (comparable with for example cultural anthropology).

In short, if theology is defined as the study of religion, there would seem to be little reason to deal separately with the question of the extent to which it is a science, and given the rather broad acceptance of the disciplines with which it is aligned, there would also be little reason to cast doubt on its academic status.[7] However, the consequence of this definition is that we systematically leave to one side the consideration of the *truth claims* which are made in the religion under investigation. For clearly, specialists in the study of religion do not consider it part of their work or competency to form a judgment about this. But the question is whether by leaving its truth claims out of the picture researchers really take the religion concerned with full seriousness, or whether in such a case they in fact avoid the question of truth.

Secondly, we can also define theology, broadly speaking, as the systematic study of the Christian tradition. Thus, in common parlance, people who by dint of their profession occupy themselves with for example the study of the Old Testament or church history are usually considered to be theologians, at least when they carry out their work in a theological institution and/or have completed a theological education themselves. If we define theology in this way, the question of whether or not it is scientific is not very critical either, because we could again refer to other disciplines. For clearly, this study of Old and New Testament, of church history, of modern-day Christianity etc. can take place in a way analogous to that in which the analysis of other traditions (of whatever type) takes place. In such research the use of methods which are also applied in current historical, literary and social-scientific research can suffice. So, in this case as well the question whether or not theological research can be deemed scientific does not involve a particular, distinct problem.[8]

7 Even Adriaanse and his co-authors do not do this; they do not in principle consider the scientific status of theology-as-the-study-of-religion as problematic. Rather, they see in this the only possibility for a scientific theology (*Het verschijnsel theologie*, 66-73, 80ff.).

8 We can of course though ask whether the existence of separate theological *faculties* does not become problematic with this definition of theology. Out of pragmatic considerations we could equally well spread theological research across the different faculties to which the different sub-disciplines are methodologically related. In fact, this is what regularly happens nowadays when theological faculties (or other theological institutions) are being restructured and embedded in more comprehensive departments of a university.

However, in regard to this (widely accepted) understanding of theology the question arises as to whether we take the bible and the Christian tradition completely seriously if we do not want to occupy ourselves with what is at stake in both of these at the deepest level, namely the truth claims concerning God and God's relation to our world. Hendrikus Berkhof has put this question succinctly to those who, for example, do want to discuss in a scientific way the *concept* of God held by Isaiah and others, but do not think they can say anything about their *God.* He asks them:

> [C]an you in a scientifically responsible way (…) look Isaiah, Paul and Martin Luther in the eyes? And above all: can you look at where their eyes are looking? For then you cut them in two; you abstract them from the one thing that passionately interests them: that in their experience they have come across no less a being than God. And in this way you abstract them from themselves. Many deform them into a sort of professor, and so into dealers in concepts. (…) they are then stripped of their threat and that is reassuring. But we have lost their object from sight, one which should also be ours.[9]

Berkhof's formulation here is rather strong (not every exegete is also a systematic theologian or has to be one), but he is in principle right.

Therefore, in what follows we too want to think of theology in a way that does not systematically exclude the question of truth. For that reason we will take theology in the literal sense of formulating a *logos* about *theos*, which may be translated as 'speaking responsibly about God'. This is precisely the classical meaning in which *theologia* was taught and studied for centuries as a discipline – in the Middle Ages even as 'queen of the sciences' – in all types of university. This concept of theology is closely related to the notion of *belief* in God that we shall also use in what follows. Some people see all sorts of critical differences between the two, and difference there certainly is, but that must not be over-stated: theology as we mean it here is the systematic reflection on, or, for that matter, the rational reconstruction of, belief in God.[10] We still find theology in this sense of the word at

9 Hendrikus Berkhof, 'God: voorwerp van wetenschap? II', in: idem, *Bruggen en bruggenhoofden*, Nijkerk 1981, 214. Similar arguments can be found in the work of adherents of the so-called 'theological interpretation' of the Bible – the phrase 'theological interpetation' indicating a way of reading biblical texts that consciously seeks to do justice to the perceived theological (i.e. God-centred) nature of these texts. Cf. e.g. Kevin Vanhoozer (ed.), *Dictionary for Theological Interpretation of the Bible*, Grand Rapids 2005, and D. Christopher Spinks, *The Bible and the Crisis of Meaning. Debates on the Theological Interpretation of Scripture*, London n.y. [2007], 7 and passim.

10 For a similar definition, see Nancey Murphy, *Theology in the Age of Scientific Reasoning*, 196.

work in disciplines such as practical theology[11] and theological ethics[12], but probably most explicitly and clearly in dogmatics. It is *Christian* theology in particular that we shall thus focus our discussion on. Much of what will be said, however, also applies *mutatis mutandis* for theology in other religious traditions, for the central question we are dealing with is a most general one: can theology in the sense of 'speaking responsibly about God' be scientific in any sense?

This question is substantially more interesting here than in relation to the previous definitions of theology that we gave. For clearly, we are now going to stop beating about the bush and talk about God directly. Therefore, we write this word here just like that – without quotation marks or any other sort of warning signs. The question of whether this sort of speaking-about-God-just-like-that is scientific or not, no matter how systematic it is, seems in the first instance to be extremely problematic. Are we not in fact starting from a pre-supposition – namely that God exists – which is not shared by everyone by a long shot? Moreover, is this pre-supposition not totally determined by our view of life, and so unproveable? How could we test such talk of God at all? Add to this the view that everyone understands his or her own thing by 'the little word God', so that its meaning is definitely not clear in advance, and the problems seem to be insurmountable. To what extent this is indeed the case, however, depends as we have said to a large extent on what exactly we mean by 'scientific'. For this reason, in what follows we shall look at the relationship between science and theology according to each of the five approaches of the philosophy of science that we have discussed. In doing so, we shall concentrate in particular on theology in the classical sense of the word (but not always mention this in what follows), taking as our text-book example dogmatics. So we will not make things too easy by taking refuge in a more palatable definition of theology.

Secondly, we want to look carefully now at the question of whether theology should *want* to be scientific at all. For clearly, we can have a separate discussion about this. Often the question is answered in the negative, not least by theologians themselves. Some, as we shall see in Chapter 7, even view making a link between faith and science as one of the most stubborn and serious mis-understandings which has arisen for centuries in this area. As to theology in this connection, it is either coupled positively with faith, or negatively with science. In the case of the former, people usually associate theology with art, with poetry, possibly also with mysticism or just with practical everyday wisdom – but in any case with something very much different from science. Theology concerns itself with things that are perhaps true in a very profound way, but not in the reality-depicting way in which scientific

11 See F. G. Immink, *Faith. A Practical Theological Reconstruction*, Grand Rapids 2005.

12 See Micheal Banner & Alan Torrance (eds.), *The Doctrine of God and Theological Ethics*, London 2006.

statements are true.[13] In the latter case, if faith is placed on the correct side of the line and theology along with science on the wrong side, then the term 'theology' only sets alarm bells ringing. Are not theologians masters in the deformation and falsification of pure faith with their (pseudo-)scientific approach?

First of all, we need to make clear here that the question of whether theology should want to be a science is a different one from the question of whether it can be so. If we define the relationship between both questions somewhat more precisely, then we have to say that the question of whether theology can be a science precedes that of whether it also should want to be so. For clearly, if it is denied that theology can harbour pretensions to be scientific, then of course it should not want to be so either. Alternatively, if it is not impossible to consider theology as a science, then it does not follow that this must therefore necessarily happen. In this sense the results of our exercise remain limited, because they remain open to different interpretations. However, in practice there will be few who think that although theology can be carried out in a scientific manner, nevertheless we should not do so because it would be inappropriate. The reason why people do not consider theology as a science is usually simply that they think that it cannot be carried out in a scientific manner. However, this is exactly what we want to examine.

And here we come to a second consideration. Faith traditions form particularly complex wholes. They comprise rites, stories, moral convictions, practical advice for living well, religious experiences, eschatological expectations, guiding ideals et cetera. But they also without exception include *cognitive beliefs* concerning the reality within which we live. Such beliefs in the Christian tradition are for example that there is a God, that Jesus Christ died on the cross for our sins et cetera. Such convictions can be seen as peripheral in relation to what faith is about at the deepest level (for example about mystical experience, or about behaving correctly, or about passing down stories from generation to generation – all these are live options). But it cannot be denied that this sort of conviction about what is really the case is an integral part of faith. In fact, faith becomes something else if we try as it were to filter out all cognitive beliefs from it.

It is exactly when it comes to these convictions about what is or is not the case, that faith has at least a point of contact with science. For clearly, science concerns itself in the broadest possible sense of the word with the reality that surrounds us. It is true that in doing so it limits itself methodologically to the empirically observable reality, but believers will in general claim that their cog-

13 This view of theology is found for example in so-called 'non-realism'. See Colin Crowder (ed.), *God and Reality. Essays on Christian Non-Realism*, London 1997.

nitive faith convictions (or 'constatives'[14], or ontological implications, or referential statements, or factual assertions, or whatever you want to call them) are also somehow related to this empirical reality. Otherwise their faith would be practically irrelevant. Their faith convictions would then be gratuitous and speculative, for they would have no consequences whatsoever for our lives, neither today nor in the future. But the fact is that in faith it is a matter by definition of what affects our reality and therefore of what is of importance for our situation in life. Faith is not usually too concerned about what is not in one way or another existentially relevant – there is of course nothing at stake in whether one believes in existentially irrelevant things or not.

Our third consideration is as follows. Theology is concerned amongst other things with reflecting on the significance of faith convictions, and with their relation to other convictions that people have. How for example does the belief that God stands as the Creator at the beginning of the cosmos relate to cosmological theories and convictions about a Big Bang? Suppose that from reflections such as this the following should be concluded: in the light of cosmological views about the origins of the cosmos, it is scientifically irresponsible to believe in a divine Creator. Could believers simply put such a conclusion to one side? Could they say: we know that it cannot be so from a scientific perspective, but we do not have to take any notice of that, and will continue simply to believe in a divine Creator? I do not think so. Not only would they lose credibility amongst many people by doing this, but they would also from an intellectual point of view be being dishonest. This would all the more be the case if they accepted the results of scientific research in all sorts of other areas just like that.

Bert Loonstra provides a clear account of this point:

> If we accept scientific results and their technical application in all sorts of areas, it is then hypocritical, if when it comes to the Bible and the old belief in God, we suddenly approach science super-sceptically. We should then have to live like the Amish in America, who couple their rejection of modern science with the rejection of all technical achievements that have come out of science. As long as we do not do this (and quite rightly so), we shall have to give serious attention to scientific conclusions and consider their truth content.[15]

The last point (as Loonstra also explicitly indicate) in no way means that we should just swallow hook, line and sinker the most recent theories in all sorts of areas of science, including their pre-suppositions. But it does mean that it is good

14 See for this concept J.L. Austin's classic *How to do Things with Words*, Oxford 1962; cf. Vincent Brümmer, *Theology and Philosophical Inquiry. An Introduction*, London 1981, 17f. and passim.

15 Bert Loonstra, *God schrijft geschiedenis* ['God Writing History'], 19f.

to look for something like a scientifically responsible account of the content of faith, or of the cognitive convictions that belong to faith. In so doing it can of course become apparent that believers are not operating dishonestly in their 'ethics of belief'. It is the task of theology to articulate the intellectual integrity of faith in this way. If we define theology as 'speaking responsibly about God' then 'responsibly' also means: scientifically responsibly. In this sense theology aims at giving a scientific account of faith (and what exactly this word 'scientific' means is what we shall consider in the rest of this chapter). For this reason theology should not run away from attempting to be scientific, in the sense of: speaking in a scientifically responsible way. For the intellectual integrity of faith is what is at stake here.

Perhaps we are stating the obvious here by adding that this does not of course mean that faith itself ought to 'scienticize'. There are of course many things (beliefs, actions etc.) which can be accounted for scientifically without them necessarily thereby being or becoming scientific.[16] So too faith is not a sort of scientific theory or hypothesis; it is even, as we shall see, a serious distortion to imagine it as such. For clearly, such an idea is at odds with the special character of faith as a personal dedication from an existentially experienced trust. To be able to believe, we ought to become like a child, Jesus said.[17] At the same time we already expect from children to strive for honesty. Believers who are not honest with themselves or with others in respect to what they hold as true therefore have a problem. For this reason, it is this honesty, this intellectual integrity, which theology has in mind and which is also the reason why theology must not shrink from wanting to be scientific. In short, theology is directed at giving a scientifically responsible account of faith, which does not, of course, transform faith itself into a scientific theory.

Now if we accept that theology conceived of as speaking-responsibly-about-God should indeed want to be scientific, who or what then determines whether it can be? This was the third preliminary question that we want to ask here. In this book we are taking a close look at the philosophy of science, the discipline that among other things deals with the question of whether something may be called scientific or not. The philosophy of science is, as we discovered in the previous chapters, very much in flux and in the process of development. If we look at it in more detail, at one moment theology might not, then it might, and in the next

16 To give just two examples: The belief that I had breakfast yesterday is not per se a scientific belief (like the belief that $e=mc^2$), but can very well be justified scientifically, for example if we have reason to believe that our memories are functioning properly. And putting on a pair of spectacles is not a scientific action, but can in certain cases also be scientifically responsible.

17 See for example Mark 10:15.

phase again it might not, be seen as scientific. Should we then make ourselves so dependent on it?

At this point, we have to discuss the powerful answer given by Karl Barth. In the opening pages of the first volume of his *Church Dogmatics*, Barth describes theology as a function of the Christian church. The church is the community of people that confesses God and talks about God in a distinctive way, deriving its speech from God's revealing address to humankind in Jesus Christ. Talking about God is a human activity, however, and therefore always fallible and open to misapprehension. It is for this reason that the church needs theology, as a form of critical self-examination with respect to the content of its distinctive talk about God. Its task is to examine whether the church's talk about God is still in accordance with God's gracious revelation in Jesus Christ that constitutes its being as the church.[18]

To be sure, there is no need for a special guild of theologians in this connection. The task of measuring the church's talk about God could also be fulfilled by Christian philosophers, historians, sociologians, psychologists or paedagogues. All these branches of scholarly knowledge do not have to be secular, and theology does not have an object of inquiry which is necessarily concealed from the other sciences. It was only because the other sciences did not adopt this special task that theology came in. Of course there always has been much scientific criticism of the church's message, but not in light of its origin and being in Jesus Christ. Even when they are sympathetic to the church, historians, philosophers et cetera, are usually inclined to judge the church's message in terms of principles derived from their sciences rather than in terms of the church's own principle: God's revelation in Jesus Christ – thereby unavoidably distorting and confusing the distinctive character of the church's talk about God. The separate existence of theology is an 'emergency measure' taken by the church to address this situation by organizing a more proper form of critical examination of its message.[19]

So in fact theology has a practical basis, not an epistemological one. Therefore, theology cannot and should not claim a place of its own in the system of sciences. Whether or not it should be considered or consider itself as a science is largely irrelevant, since there are both similarities and dissimilarities between theology and the (other) sciences. It seems better to refrain from this claim, however, since otherwise theology might easily be disturbed in its task by regard for what counts as science elsewhere. In fact, this is a serious danger for theology, for

> … to the discharge of its own task it must absolutely subordinate and if necessary sacrifice all concern for what is called science elsewhere. (…) As regards method, it has

18 Karl Barth, *The Doctrine of the Word of God* (Church Dogmatics I,1), 2nd edition, Edinburgh 1975, 3-4.

19 Barth, *Doctrine of the Word of God*, 5-7.

> nothing to learn from them. It does not have to justify itself before them, least of all by submitting to the demands of a concept of science which accidentally or not claims general validity.[20]

Neither should theology attempt to contribute to a better (more comprehensive etc.) understanding of science by which it is included rather than excluded. For in doing so it would again suggest that its existence is necessary in principle, which according to Barth is not the case. Theology is not to be seen 'as a link in an ordered cosmos, but only as a stop-gap in a disordered cosmos'.[21] It can only prove its scientific rigour by passionately devoting itself to its task, thus showing what it means by true science. For clearly, what may count as a science is not decided upon by conventions and stipulations regarding a general concept of science, but by what is being done in practice.[22]

In this final statement we may hear an early prelude (Barth wrote this in 1932) to the Kuhnian shift from a prescriptive towards a descriptive theory of science.[23] So with the benefit of hindsight we can say that Barth was definitely right here. But what about his overall position on the status of theology? It seems to me that Barth is right that theology can only start from revelation. As Pannenberg rightly says, human knowledge of God can be true knowledge 'only if it originates in the deity itself.'[24] It is also true that according to Christianity God's revelation found its culmination in the life, death and resurrection of Jesus Christ, so that the Christ-event is constitutive for the Christian concept of revelation. However, it seems to me that Christian theologians cannot simply start their work at this point. If they are to be accountable to everyone who has a concern for the object of theology (as Barth concedes)[25], they should be able to explain why they are Christians in the first place, i.e. why they hold the Christ-event to be the constitutive revelatory disclosure of God. And it seems that this can only adequately be done in generally accessible language, i.e. by making use of generally understandable criteria for truth and knowledge, rules of logic et cetera. For if in the end theologians retreat to an appeal to purely subjective convictions and idiosyncratic conventions, their attempt to give an account to anyone who asks them 'a reason for the hope

20 Barth, *Doctrine of the Word of God*, 8.

21 Barth, *Doctrine of the Word of God*, 10. Ideally, Barth suggests, everyone would know God in an intimate way, so that no theology is needed (cf. the reference to Jeremiah 31:34 on page 5).

22 Barth, *Doctrine of the Word of God*, 10. For a comprehensive elaboration of Barth's view of theology, see Hermann Diem, *Theologie als kirchliche Wissenschaft*, three volumes, Munich 1951-1963.

23 See above, section 2.4.

24 Wolfhart Pannenberg, *Systematic Theology* 1, Edinburgh 1991, 189.

25 Barth, *Doctrine of the Word of God*, 8.

that is within you' (1 Peter 3:15) threatens to become barren, and theology itself unduly insulates itself from the wider world.[26]

Does this mean then that theology must adopt the criteria for scientifically reliable knowledge as expounded in contemporary philosophy of science? Not necessarily so. We can at least learn from Barth that a certain amount of scepticism is justified here. This is just as much the case when the philosophy of science excludes the possibility of a scientifically responsible speaking about God as when instead it allows for this possibility. When the philosophy of science does exclude it, we should not let it lead us uncritically. It is better to ask *on what grounds* it makes this exclusion.[27] But then we shall have to take these grounds very seriously and have to consider how we can deal with them. Perhaps it is possible to make plausible that theology (for example by dint of the special character of its object) has its own particular contribution to make to this discussion. Perhaps the legitimacy of the reasons adduced can be doubted on other grounds. However that may be, although a critical attitude is appropriate, this does not mean that we can easily dismiss what the philosophy of science contributes to a discussion on the status of theology. If those who argue that theology in the classical sense of the word can make no claim to be scientific are right, then we should at least be worried by this. For this would imply that the truth claims of faith cannot be accounted for in a normal way.

How would things stand, however, if the philosophy of science were to open up the possibility of a scientific theology, and even perhaps to provide important reasons for this? In that case as well we would need to operate very carefully. If for example science were to be described in a relativistic way as an entirely irrational matter (as we saw happen with Feyerabend), then it is doubtful whether theology gains much by appealing to this development in the philosophy of science.[28] A

26 Barth, *Doctrine of the Word of God*, 8. Cf. for similar criticism, issuing from a survey of Barth's discussion with philosopher and logician Heinrich Scholz on the scientific status of theology, Pannenberg, *Theology and the Philosophy of Science*, Philadelphia 1976, 265-276.

27 I have been told that at one time a short discussion took place between two philosophers of religion, the first of whom argued that if only 10% of what Martin Heidegger had written were true, the prospects for Christian faith would be very dim. His opponent, however, quite appropriately answered him by asking: 'But *which* 10% of what Heidegger has written is true, and on what grounds do you think so?' To that question, there came no clear answer ….

28 It is interesting that Feyerabend himself, according to personal correspondence with his pupil Nancey Murphy (16-1-1986), considered it an important 'propaganda-stunt' (thanks to the great respect that science still enjoys in Western society) if by appealing to his definition of science we were to claim that theology can satisfy the demands of being scientific. See Murphy, *Theology in the Age of Scientific Reasoning*, 84.

theology that situates itself all too eagerly in the hole that post-modernism has cut in the hard ice of modernity can as it were all too easily drown. In what follows, we shall indeed see (as we would expect to on the basis of the previous chapters) that the chances of a scientific theology are much more favourable in the present climate of the philosophy of science than they were in the positivistic one of half a century ago. On the one hand, this development should not be acclaimed uncritically, since if it is not based on good reasons it may soon evaporate. On the other hand, it is also a matter of intellectual honesty to take it seriously, study its backgrounds, its consequences, et cetera. It is precisely this what we intend to do here.

If it turned out that to speak in a scientifically responsible way about God is indeed very well possible given the current state of affairs in the philosophy of science, then this does not mean that theologians should immediately (and perhaps even triumphantly) consider themselves rich as a result. But at least it does mean, put very simply, that the opposite claim has no warrant for the time being. In other words: it means that it is incorrect to assert that talking about God these days could not or could no longer take place in a scientifically responsible way. And this is in fact the one thing that we will demonstrate in what follows. If we manage to do this, then the burden of proof will shift back onto those who assert that theology is not scientific, that it does not belong in the secular university, and that it should also not play a role in public debate.[29]

5.3 Theology in a positivistic climate

In this section, we shall discuss the consequences of the three currents in the philosophy of science that we described first in the introductory paragraph, and which from a historical perspective were also in fact the first to come about: the original and the revised form of logical positivism, together with critical rationalism. It is true that the third of these departed on essential points from logical positivism, but it still continued – certainly with hindsight – to a large extent to breathe the same air. So, for example, it still aimed at bringing about a sharp dividing-line between science and everything that incorrectly assumed this title of honour. In this sense, with critical rationalism we continue operating in what we could call a positivistic climate. In such a climate, what are the chances for a scientific theology?

If we begin with logical positivism, then there is a short, sharp answer: nil. Alfred Ayer in his time needed only a few pages to account for this answer.[30] Statements about a transcendent God possibly say something about the state of mind of the speaker. But they do not contain any factual information. For clearly, if a statement

29 So correctly Murphy, *ibid.*, 85.
30 Ayer, *Language, Truth and Logic*, 119-126.

contains any factual information, then we must be able to say how this can be empirically verified. Now in the case of God this is impossible, because by definition God is considered by believers not as an empirical entity but as a transcendent one, which 'is held to be superior to the empirical world, and so outside it'.[31] But then sentences in which the word 'God' occurs are not empirically verifiable. And so they have no factual content. It is worth noting that in fact Ayer here associates himself with what believers say. For usually, believers admit that their religious truth claims cannot be tested in any empirical way. For example, we cannot carry out any experiments from which we can necessarily deduce the existence and activity of God, as the priests of Baal must have recognized in their competition with the prophet Elijah.[32] It belongs to the very deity of God, as believers rightly say, that God surpasses human understanding and does not put Godself at their disposal. As a result of this *Unverfügbarkeit Gottes* (unavailability of God), statements about God are not verifiable indeed. As a result, according to the logical positivistic criterion they have no meaning.

This is the case not only for primary statements of faith, but also for the more systematic, ordered statements about God such as those which are made in theology. For this reason these cannot be reckoned to the field of science either. Theological statements are in fact text-book examples of unproveable metaphysical statements, or to put it less politely: of hot air .... According to Rudolf Carnap, the logical-positivistic verification procedure had demonstrated that traditional theology is a relic of earlier times, totally alien to the scientific way of thinking of the twentieth century.[33]

The logical positivistic procedure for testing the scientific content of statements seemed to be as simple as it was effective, and it was very influential in particular in the 1950s and 1960s. Theologians felt forced onto the defensive and tried in different ways to make a virtue out of a necessity. According to some, such as Paul van Buren and Richard Braithwaite, Christian faith was to be seen not at all to be about factual assertions concerning transcendent metaphysical entities, but rather about articulating a certain moral outlook on life, which goes back to the historical Jesus. In accordance with this view, they argued that the Gospels could be translated without any significant loss in terms of this moral outlook on life.[34]

31 Ayer, *Language, Truth and Logic*, 122.

32 For this Old Testament story, see I Kings 18. Even Elijah could not of course have God 'at his disposal'.

33 See Schlipp, *The Philosophy of Rudolf Carnap*, 8.

34 Paul M. van Buren, *The Secular Meaning of the Gospel*, London 1963; Richard B. Braithwaite, 'An Empiricist's View of the Nature of Religious Belief', in: R. B. Braithwaite (ed.), *The Philosophy of Religion*, Oxford 1971, 72-91. For a good discussion (which takes into account the mutual differences between them) as well as a convincing rejection of these views, see Vincent Brümmer, *Theology and Philosophical Inquiry*, 257-268.

Others admitted that the logical positivists were right concerning the here and now, but not necessarily concerning the future. They discussed the possibility of an 'eschatological verification' of Christian belief statements. So, for example, I. M. Crombie and John Hick asserted that religious and theological statements are in fact in principle verifiable, but that such a verification (or falsification) will not practically speaking be able to take place until the eschaton. For clearly, only then shall we have the total overview. The end of history will then for once and for all determine whether or not believers are right ... Hick compared the situation with that of two travellers, who are on the same road and who have the same experiences. According to one, the road led to a certain destination, according to the other this was not so. The former interpreted his experiences in accordance with his belief in that destination, whilst the latter explained everything that happened on the journey in another way. Which of the different interpretations was true could not be determined on the journey itself. But this still does not mean that an interpretative statement of the believer such as 'my life is a pilgrimage on the way to the Kingdom of God' is meaningless. For the possibility should not be excluded that this statement will be verified down the line.[35]

In this way Hick shed light on the deficiency of the verificationist criterion of meaning: it will not do to make the factual content of an assertion dependent on the practical verifiability of it. A believer who expresses her faith in a transcendent God certainly does not by definition speak nonsense in doing this: although it cannot be verified right now, her belief might turn out to be true in the future. Still, however, it is not clear how on the basis of this sort of belief statement of faith a scientific theology is possible. For clearly, in science we do not get away with a reference to the possibility of an eschatological verification. On the contrary, by using verification procedures that can actually be applied, it has to be determined here and now which statements may be considered as scientifically sound and which not. As long as this remains impossible in relation to religious statements, it is true that such statements do not have to be meaningless (in the sense of cognitively empty), but no scientific theology can be developed on the basis of them. For this we simply need the possibility of verification here and now. In short, theology cannot be carried out as a scientific activity on the basis of the fundamental principles of the original logical positivism.

We have seen, however, that this original logical positivism did not hold its position for very long (although it did carry on exerting a huge influence for a

35 I. M. Crombie, 'Theology and Falsification', in: A. Flew & A. MacIntyre, *New Essays in Philosophical Theology*, London 1955, 109-130; John Hick, 'Theology and Verification', *Theology Today* 17.1 (1960); developed further in John Hick (ed.), *The Existence of God*, London 1964, 252-274.

long time), but was succeeded by the revised variant that we described above as *confirmationism*. So what are the chances of a scientific theology on the basis of the confirmationist criterion for being scientific? In any case, much greater. According to confirmationism, strict verification is not necessary for a proposition to be scientifically warranted; it suffices if we find enough confirmation for it. Now it is interesting to note how much confirmation there is for all sorts of belief statements. So, for example, there are millions of people who say that they have experienced God in their lives (in any case they far outnumber those who for example have seen UFOs). Some of them even claim that they experience God's presence every day. Of course, it is possible that people may be mistaken in this. But is it not remarkable that so many, who in addition are to a great extent independent of each other and spread across so many times and places, make this sort of statement, and still maintain their position when it is subjected to critical examination? In short, the statement that God exists has a particularly high degree of confirmation, and can within a confirmationist framework therefore be considered as scientifically tested (and therefore as true).

In this regard it is useful to point to a remarkable phenomenon which logicians have been confronted with for a long time now, namely the fact that so-called existential propositions are much easier to confirm than non-existential propositions. In order to confirm a statement such as 'rhinoceroses exist', it is for example only necessary that someone observes a minimum of two rhinoceroses. Especially, if several people quite separate from each other claimed that they had observed rhinoceroses under given circumstances, then there would be hardly any further reason to doubt the existence of such animals, even for someone who has personally never come across one. By contrast it is much more difficult to support a non-existential proposition such as 'no rhinoceroses exist'. For where do you begin and finish such a process? Even if we search the whole world square metre by square metre and do not come across a rhinoceros, it still will not have been demonstrated that they do not exist. For of course it cannot be discounted that they have changed places in the meantime. Thus, existential propositions are always more difficult to confirm because one counter-example is enough to undermine them – and such a counter-example can always pop up anywhere.

Similarly, in a confirmationist framework it is so much more complicated to demonstrate the non-existence of God than his existence. If only one person says that she has encountered God and there is no reason to for thinking that she is mistaken (she is not stupid, not gullible, not ill, does not make mistakes very often etc.), then already it seems natural to accept that her story is true, and that God does in fact exist and can be encountered. If someone else by contrast says that they have not met God, this does not say anything at all about whether or not God exists. For clearly, there are many living beings that we have never met, but which certainly do exist.[36] Now it could be replied that in the case of a rhinoceros the person who has encountered it must be able to show where this happened. And then using the tracks it has left behind we can try to discover the rhinoceros, so that others too can be certain of its existence. In this case we indeed have a cumulative confirmation for the thesis that rhinoceroses do exist. Is not the problem then with statements of faith such as 'I have experienced God' the fact that they do not allow for others to share in this experience, by being taken as it were to its source? As a result, it seems that we shall simply have to take their word for granted.

That remains to be seen, however. William Alston, who carried out much research on the epistemic status of alleged human experience of God, shows that such experiences usually only take place within a certain context, namely the context of faith, prayer, susceptibility towards God's word (for example through the Bible) and other attitudes and activities that belong to a spiritual life. Anyone who does not seriously want to make such a context his own, must not complain about being excluded from God-experiences,[37] just as someone who does not want to leave their home cannot complain about the fact that they have never come across a rhinoceros. The reverse is perhaps just as true: anyone who strives to find God in an appropriate way will find God sooner or later. So it is questionable whether both situations are structurally so different. For all sorts of people and other beings it is the case that you have to look in the right way and in the right places in order to be able to find them.

But even if we have a relevant difference here, the question still is whether it is decisive. For clearly, there are many people who appear so trustworthy to us that generally speaking we take them at their word without giving it a further thought, even when we ourselves do not have the experiences that they report. Why would we then depart from that principle whenever such people assert that they have encountered God? And although we are not ourselves in a position to check their assertions, many others confirm them for us. In a similar manner we believe a whole raft of statements made by natural scientists. We cannot usually personally check what they say about light years or electrons, but we consider that they are trustworthy people and we know that others who find themselves in the appropriate circumstances (i.e. colleagues of theirs) confirm their findings. Likewise there is a lot of confirmation of the reality of encounters with God.

36 For the meaning of the word 'exist' in this connection, see Brümmer, *Theology and Philosophical Inquiry*, 268-282.

37 William P. Alston, 'Christian Experience and Christian Belief', in: A. Plantinga & N. Wolterstorff (eds.), *Faith and Rationality*, 129. See also the further development of this in Alston, *Perceiving God. The Epistemology of Religious Experience*, Ithaca 1991.

Now it might be objected that there are clear differences between empirical evidence as based on sensory experience in the form of observations and experiments on the one hand and religious experience on the other. Only the first of these types of experience – very broadly speaking: sensory experience – counts within the (revised) logical positivism as confirmation. Religious and other sorts of non-sensory experiences (such as moral or aesthetic) cannot operate like this as confirmative with regard to the truth, probability or scientific nature of propositions. But then we are faced with the task of finding out why this should be so, or why the differences between sensory experience on the one hand and for example religious experience on the other is so important in this regard. In response to the research of William Alston that we have mentioned, as well as to recent succesful attempts to link religious experiences just like sensory ones to certain specific parts of the brain, a renewed debate on these issues has arisen.[38]

Imagine that it is indeed reasonable to ascribe to religious experiences a degree of confirmation with regard to religious statements as high as that ascribed to sensory experiences with regard to statements about the sublunary world. Then a scientific theology based on confirmationist principles would be entirely conceivable. This would then have as its task not only comparing different religious truth claims with each other and testing them for consistency, coherence et cetera, but also evaluating the experiences by which they seem to be confirmed. Moreover, it should be examined how it is possible that these experiences are so varied in nature (whilst people's sensory experiences usually in fact agree with each other to a large extent). Experiences of God or the divine are as is well-known interpreted very differently not only in different religions, but also within one and the same religious tradition. We should track down the factors which play a role here in the hope of being able to purify the appeal to experience from possible errors. In this way a sort of natural or philosophical theology based on religious experiences should be able to take shape (and is in fact taking shape as for example with Alston).[39] To be able to take part in such a theology no prior belief is required, but only the readiness to take religious experiences as completely seriously as sensory ones. The matter is in any case important enough, for it may make quite a difference to our life and future whether there is in fact a God who can be encountered or not!

However, we have seen that confirmationism too was not blessed with a long life in the philosophy of science, because in the course of time it was overtaken by

38 See e.g. Andrew Newberg, Eugene d'Aquili & Vincent Rause, *Why God Won't Go Away. Brain Science and the Biology of Belief*, New York 2001.

39 Which is not to say that Alston operates according to the classical principle of confirmation; he takes opposition to his theory too seriously for this to be so.

Karl Popper's critical rationalism. According to Popper, from a scientific perspective there is nothing to be gained from only looking for confirmation for one's hypotheses and theories, because in principle any discovery or any relevant event can be explained as such a confirmation. Often not much creativity is necessary for this – only that much more gullibility, for in this way people make themselves immune to all possible criticism in the form of serious voices of opposition. According to Popper's falsificationism, however, it is precisely these 'voices of opposition' (or to put it more accurately: falsifying observations) that we should look for. And it is only by being confronted with these that we can see whether the theory can really hold its own.

Arguably, this basic insight of Popper's is also of importance for theology. Anyone who develops a theological theory, i.e. who makes a series of connected assertions about God and God's relation to reality, will not simply have to make mention of only those views and considerations that are intended to confirm this theory. For then theology will quickly become ideological like Adlerian psychoanalysis and classical Marxism. Data that do not fit the pattern are then not taken seriously but interpreted in an artificial manner and forced into the straightjacket of the theory. Therefore, anyone who brings forward a theological theory will do well to confront it explicitly and honestly with the counter-arguments. And so the alternative theories ought not to be made as weak as possible but rather as strong as possible. If it can then be shown that the alternatives are problematic, serious development takes place in research. This is still not per se the same as progress, for later in a different light the presumed gain can turn out to be a loss, but it is however the sort of step forward that may be expected from scientific research.

So here we should continue to learn from Popper, all the more so because in all sorts theological research his critique of inductivism (i.e. of the search for ever more different observations which should confirm a theory) is not always sufficiently taken into account. Theologians sometimes write impassioned arguments in which they set out a definite position, but in doing so cut themselves off completely from what critics of this position have argued, instead of taking these critics as seriously as possible. If these critics are being referred to at all, it is often in a disqualifying way, as if they have simply understood nothing of what the theory is about. But more often still the critics, who usually live and work within another theological 'paradigm', are simply ignored.[40]

40 The way in which Dutch theologian H.M. Kuitert develops his christology is a case in point. See his *Jesus: The Legacy of Christianity*, London 2000; cf. my comments in Gijsbert van den Brink, *Een publieke zaak*, Zoetermeer 2004, 221-231, and also the critical British response in *Expository Times* 111 (2000), 397f.

Let us now, however, turn to the critical question which we must ask in connection with Popper's critical rationalism as we did in relation to logical positivism: given its (falsificationist) criterion for science, are there possibilities for a scientific theology, and if so, what does such a theology look like? As favourable as the wind previously was for confirmationism, so now it is equally unfavourable. For what sort of falsifiable predictions would be possible in faith and/or theology, that can also be tested in practice? We remember all too well how movements such as the Jehovah's Witnesses made attempts in the past to predict the date of the return of Christ. These were in any case predictions that could be tested! The dates predicted, however, passed without anything remarkable happening – but also, and this is just as interesting, without the membership of the religious movement in question being decimated. Clearly, the non-fulfilment of the prediction made by the leaders was not considered as a crucial experiment in the Popperian sense, and these leaders for the most part continued to be trusted.

The question then arises of whether there are in fact any future events at all imaginable in which believers would see a reason to give up their faith. In other words: can believers make themselves vulnerable as the Popperian scientist ought to make himself vulnerable, by admitting: 'If under circumstances x and y event z takes place, then I shall renounce my faith and recognize that I got the wrong end of the stick'? When Popper's ideas became influential, it occurred to the British philosopher Antony Flew, who was one of the world's foremost philosophical atheists at the time[41], that believers never in fact speak in this way.

Now often it seems to people who are not religious as if there was no conceivable event or series of events the occurrence of which would be admitted by sophisticated religious people to be a sufficient reason for conceding "There wasn't a God after all" of „God does not really love us as a father loves us then." Someone tells us that God loves us as a father loves his children. We are reassured. But then we see a child dying of inoperable cancer of the throat. His earthly father is driven frantic in his efforts to help, but his Heavenly Father reveals no obvious sign of concern. Some qualification is made – God's love is "not merely a human love" or it's "an inscrutable love," perhaps (…). I therefore put (...) the simple central questions, "What would have to occur or to have occurred to constitute for you a disproof of the love of, or existence of, God?"[42]

41 Half a century later, Flew would exchange his atheism for a form of theism; see Antony Flew & Gary Habermas, 'My Pilgrimage from Atheism to Theism. A Discussion between Antony Flew and Gary Habermas', *Philosophia Christi* 6 (2004), 197-211. Since this article was published, on the internet the debate about the background and sincerity of Flew's 'conversion', on the present state of his mental capacities etc. is going on and on …

42 A. Flew, 'Theology and Falsification', in Flew & MacIntyre (eds.), *New Essays*, 98f.

These questions sparked off a lot of reactions and quickly became known as 'Flew's challenge'. Flew illustrated his challenge in a rhetorically brilliant way by means of a re-telling of a parable from John Wisdom. This concerns two explorers who come across a clearing somewhere in the jungle where many flowers and weeds are growing. According to one of them there has to be a gardener who has tends this plot. According to the other this is not the case. They decide to keep watch over the clearing, but they never see a gardener. So, the first one suggests that there is possibly an invisible gardener. Both of them decide to set up an electrified barbed-wire fence around the clearing and also to patrol the area with bloodhounds. But there is no movement in the fence, no scream which tells them that someone is trying to get through, no sign of danger from the dogs. But the 'believer' is still not convinced. 'I still believe that there is a gardener', he says, 'invisible, intangible, insensible to electric shocks, a gardener who has no scent and makes no sound, but who comes in secret to look after the garden that he loves'. His fellow explorer then asks somewhat bemused, what the difference is between an invisible, intangible, ever elusive gardener and no gardener at all?[43]

Flew clearly evokes the intellectual climate here that Popper challenged: whatever happens or does not happen, nothing will be able to deprive the first explorer of his theory that there is a gardener. But such a theory in fact has no content. Flew even seems to draw a conclusion that Popper did not draw: that such a theory is also meaningless (Popper would not go further than 'not scientific'). If in the same way faith in God is meaningless, then in this regard we are back with Ayer, and the possibility of a theology that speaks about God in a scientifically responsible way is ruled out in advance.

We have seen, however, that the falsificationism of Popper turned out in the final analysis to be untenable, because falsification is never a matter of a simple crucial experiment with an unambivalent outcome. According to the thesis of Duhem and Quine, when we are confronted with a falsifying observation there are always various paths that we can take to resolve the problems, and it cannot be determined in advance which one is correct. We have already seen that natural scientists often deal with falsifying observations in exactly the same way as the 'believing' explorer did with the failed attempts to track down the gardener: the theory is not rejected, but it is modified in certain ways – if necessary almost endlessly. What is interesting here is that in many cases this is a completely rational way of carrying on, because the theory in question has often already gone through a great number of tests and not for nothing has it acquired an almost sacrosanct status. This was the case for example for the Newtonian theories which were not simply rejected by the imaginary physicist in Lakatos' example when he

43 Flew, 'Theology and Falsification', 96.

could not reconcile certain observations with them. Can it not similarly be completely rational when a believer does not reject his faith in God and his love, even if at a certain moment everything seems to point away from it and nothing towards it?

Let us look somewhat more closely at the case that Flew presents in this connection. We shall broaden the example of the child that is dying from cancer of the throat to include all distressing situations of suffering that happen in our world. Given faith in a God who is as loving as he is omnipotent, the occurrence of such an unimaginably large amount of suffering is of course highly problematical. Must not believers, if they are honest, just admit that their faith in an omnipotent and loving God has been falsified already for a long time now by the facts? Indeed many draw this conclusion. However on closer inspection this turns out to be premature. The question here too is: *what exactly* is falsified by the great evil and suffering that there is in the world? In this case as well it turns out that this question cannot be answered in a single unambiguous way. We can as it were intervene at various places in the web of belief to account for experiences of evil and suffering. The conclusion that the existence of an omnipotent and good God is falsified by evil only holds if we subscribe to a number of hidden presuppositions.[44]
Amongst these are:

1) It is inconceivable that the evil in the world is permitted by God with an eye to a greater good, which without (the possibility of) that evil would not be attainable.
2) We know enough of God to be able to judge all his actions from a moral point of view.

If we are moved by the inexpressible amount of evil and suffering in the world and by the appalling seriousness of it, this might also bring us to give up 1) or 2) instead of faith in the love and omnipotence of God. If we give up 1), then we recognize that (the possibility of) evil in this world perhaps has to contribute to a higher good which is not realizable without it (not even for an omnipotent God, assuming with far and away the majority of thinkers that even an omnipotent God can not bring about what is logically impossible to bring about). Many thinkers believe that human freedom is such a higher good. If we do not want to adopt such a view, however, then we can also look critically at supposition 2). How do we in fact know that we are in a position to judge the priorities of an omnipotent and completely good God from a moral point of view? Is it not much

44 For a similar train of thought, see McGrath, *The Foundations of Dialogue in Science and Religion*, 94-98.

more probable that this God does things that are far beyond our ability to comprehend (let alone our ability to judge)?

Whatever the case, it may be clear that a simple falsification procedure is not sufficient for us to reject the 'theory' – in this case the existence of (an omnipotent and completely good) God. On the contrary, this extremely complicated problematic cannot just be resolved by means of a single 'crucial test'. We are dealing here with the effects of the thesis of Duhem and Quine: if there is a short-circuit between a theory and an observation it is theoretically impossible to say precisely what part of the theory is causing this short-circuit. The extraordinarily sophisticated debate that is taking place in contemporary philosophy of religion concerning the so-called problem of theodicy (i.e. concerning questions about God and evil), demonstrates this. Even though different positions have been taken up in this debate, the popular reasoning of 'if there were a God, then he would do more about the misery in the world', turns out to be much too simplistic and in this form quite simply unsound.

If we respond to Flew's challenge in this way, we have in fact departed from his intellectual framework, strongly informed as it still was by the older positivistic philosophy of science. For clearly, the analogue in the philosophy of science of the above reaction to Flew's challenge is to be found in holism, as this has shaped the philosophy of science ever more emphatically in particular since Quine: in the debate over whether or not something is scientifically warranted we must not use isolated statements as our starting point, but rather comprehensive networks of convictions which are connected to each other in a complex way. So it is now time to consider whether, and if so how, a scientific theology is possible on the basis of notions of science which emerge from the newer philosophy of science. But before we do this, we have to make two concluding remarks.

First of all, it is not the case that insights and convictions inspired by the older philosophy of science are no longer of any value at all. On the contrary, all sorts of elements from the classical conceptions of science (as we might call them) continue to maintain their significance. In regard to the original *logical positivism* we think of the idea that cognitively meaningful statements ought in principle to be verifiable. Even though such verification, in contrast to the old ideals, is not always possible in practice, we should be able to say under what circumstances the truth would or will come to light. When this is not possible in any way (not even eschatologically), the idea that the statement in question refers to reality is based on a misunderstanding.[45] For clearly, reality can never confirm or deny such a

45 For a similar point, see Pannenberg, *Theology and Philosophy of Science*, 334 (cf. for the possibility of verifying theological statements, 338-343). It is to be feared that especially theological statements by liberal theologians, to the extent that they no longer believe in a real eschaton, come up short on this point.

statement. In regard to *confirmationism* we note as a point of lasting significance which has not become obsolete as a result of later developments, the insight that existential statements can in principle be confirmed much more easily than non-existential ones. And as regards *falsificationism* Popper's basic instinct still stands, that in science we only move forwards if we are prepared to be 'disturbed', i.e. allow ourselves to be contradicted by researchers and authors who do not necessarily affirm us in our intellectual frameworks, but are critical of them. A theology which avoids such discussions and debates in order to wallow in its own sense of being right does not deserve the predicate 'scientific'. And even if it did not place any value on that predicate it would be a rather unfruitful enterprise, in so far as it would not dare to disseminate the truth of which it is convinced by means of an honest dialogue with critical voices.

Secondly we can ask whether in current theology these and other lessons from the older convictions are not simply appropriated in isolation from their original comprehensive frameworks. In other words, has the shift from the older to the newer philosophy of science in the practice of contemporary theological reflection not already been made? Are we not tilting at windmills if we fight against theological forms of thought that still operate from within the out-dated frameworks of the philosophy of science of verificationism, confirmationism and falsificationism? However, as is clear from the example of H.M. Kuitert's study of christology mentioned above[46], this is not the case. Unfortunately there are still theologians who operate as if nothing of substance has happened in the last fifty years in the philosophy of science. By contrast, in the next chapter we shall consider the status and tasks of theology if we approach it from the perspective of the newer philosophy of science.

46 See above, footnote 40.

6. Theology and the newer philosophy of science

6.1 Introduction

Having considered in the previous chapter the implications for the status of theology of older currents in the philosophy of science, I now want to explore the prospects and possibilities for a scientific theology from the perspective of the newer philosophy of science. This is an area in which a lot of work has still to be done. Only a few philosophically inclined theologians (and theologically inclined philosophers) have led the way by reflecting upon this relationship. One of the first to think through the significance of the more recent philosophy of science for theology was the German systematic theologian Wolfhart Pannenberg. We shall see that, already very early on, Pannenberg appropriated Kuhn's insights for theology, yet continued to cling, at a decisive point, to a by now outdated view of Popper.

Next, we shall consider a contribution to the discussion by the theologian and philosopher Nancey Murphy. A student of Paul Feyerabend, she also attempted to develop, with recourse to the ideas of Imre Lakatos, a scientifically robust theological programme. It will turn out, however, that, at a crucial point, she also fails to remove doubts about the feasibility of her programme. Finally, therefore it will be up to us to inquire more straightforwardly, starting from Kuhn, into the kind of profile a scientific theology could and should have.

6.2 Theology as science in the work of Wolfhart Pannenberg

In a number of articles, resulting in his major monograph *Theology and the Philosophy of Science*, Wolfhart Pannenberg elaborated his view of how a truly scientific theology is possible within the conditions of contemporary philosophy of science. In doing so, Pannenberg's concern is, quite emphatically, with theology as scientific discourse on God, i.e. with the classical meaning of the word 'theology', as it also appears in our study.[1] At this point, he agrees with his teacher Karl Barth, who

1 Wolfhart Pannenberg, *Theology and Philosophy of Science*, 297f. (on 'Theology as the Science of God'). Recognition is due to the English philosopher of religion, Basil Mitchell, for having pointed, at virtually the same time as Pannenberg, to the great significance of Kuhn's work for the justification of religious belief. See his *The Justification of Religious Belief*, London 1973. Since Mitchell does not apply these insights to the

had no interest in any other object of theological discourse than God himself (and everything else in its relation to God). As it happens, Pannenberg's understanding of how such discourse on God might be possible not only deviates quite significantly from that of Barth, but is also very detailed and complex. It virtually encompasses, in embryonic form, his entire theological programme as he has unfolded, adjusted and elaborated it in various directions over the years. This is not the place to offer a survey of the theological thought of Pannenberg, however fascinating that might be.[2] Our concern, here, is the manner in which Pannenberg tries to do justice, in his theological method, to insights derived from historical developments in the philosophy of science.

Pannenberg begins his major study on the relation between the philosophy of science and theology with a survey of developments in the philosophy of science, starting with logical positivism. In a manner more or less analogous to that which we used in the previous chapter, Pannenberg indicates the conclusions to be drawn for the status of theology from the main thrust of each of the different currents. Yet, at the same time, Pannenberg already shows how one current almost naturally succeeds another. Since reflection does not cease when a particular model of science is confronted with major problems (which always proves to be the case), such a model is, as it were, driven beyond its own limits. Such is Pannenberg's portrayal of the transition from early positivism to logical positivism, from logical positivism to critical rationalism, and then, albeit to a lesser extent, from critical rationalism to Kuhnian paradigm theory. Moreover, in this way he sketches the development from the monopolising of the concept of science by the natural sciences to the rise of the humanities and the hermeneutical method.[3] In every case, he discerns a kind of natural dynamic that makes it impossible to hold on to positions initially adopted, because, on closer inspection, they turn out to be inadequate in some way or other.

This does not mean that the older currents have no insights of lasting significance to offer. This is why Pannenberg traces them so carefully. Thus he sees much that is good in Popper's critical rationalism and, to a lesser extent, in logical positivism. Yet, regarding their most central distinctive characteristics, he feels bound to conclude that they have lost their validity. In his opinion, this applies both to the verification principle of logical positivism and the falsification principle of critical rationalism – at least in its strict application, as advocated by Popper.

status of theology, but limits himself to the discussion in the philosophy of religion on theism and atheism, we shall not consider his contribution here.

2 Compare, in this regard, e.g. Gunther Wenz, *Wolfhart Pannenbergs Systematische Theologie*, with a selection of secondary literature on Pannenberg's work at pp. 301-324.

3 Cf. Pannenberg, *Theology and Philosophy of Science* 27, 29-71; 72-155. Cf. also Wenz, *Wolfhart Pannenbergs Systematische Theologie*, 23-25.

Pannenberg agrees with Popper that statements can only be taken as having cognitive content, and as being scientific, if they can, in principle, be tested. The question, however, is: tested *against what*? Not against 'hard facts' or 'objective observations', for as Popper himself realised such things do not exist. Rather, they should be tested against certain observational sentences, i.e. sentences *about* particular observations, such as 'this table is white'.[4] Yet sentences of this type always turn out to contain certain convictions (e.g. 'white is a colour', or indeed, 'white is not a colour'), which are, in turn, theoretical in nature, and should therefore be subjected to critical testing themselves. But this critical testing would again yield similar problems! And indeed, Popper admitted that observational sentences can only be employed as testing criteria because and in so far as experts are so convinced of their correctness that they agree that they can serve as such. In other words, they do not rest on objective data, but on intersubjective agreement.

However, according to Pannenberg, Popper did not recognise sufficiently just how problematic this element of 'conventionalism' (testing criteria rest on conventions, or implicit agreements, rather than objective data) was for his search for a clear demarcation between science and metaphysics. After all, every community has its own conventions, and it is not possible to say in advance which one is better than others, since these conventions cannot be tested exhaustively. This means that they are, in fact, meta-physical in nature, so that science and metaphysics are caught up in one another! As a result, conversely there is no reason to judge in advance that those conventions within which divine action has an accepted place (because the underlying observational sentences refer to divine acts) are scientifically unjustifiable. Pannenberg mentions as an example here the Israelites' experience of the passage through the Red Sea – a passage which was observed to be the result of a wonderful act of God (of course, he could as well have mentioned more contemporary examples). Obviously, there is a clear 'metaphysical' element in such observational sentences – but then even natural scientists cannot escape the fact that the 'basic sentences' with reference to which they test their theories are also ultimately rooted in worldviews of a religious or (more probably) philosophical nature.[5]

4 Carnap spoke, in this regard, of 'Beobachtungssätzen' (observation sentences), Neurath of 'Protokollsätzen' (protocol sentences), and Popper of 'Basissätzen' (basic sentences).

5 Pannenberg, *Theology and Philosophy of Science*, 54f., cf. 50-58 as a whole. 'Not only does the language required for the formulation of basic propositions root natural science in the psychology of the individual investigator, but the logic of its statements fixes it firmly in general attitudes which are ultimately philosophical or religious in character' (55).

At this juncture, Pannenberg then links up with Kuhn, according to whom, as we have seen, testing procedures can only take place within such larger theoretical wholes, so that what is ultimately at stake is the credibility of these larger wholes (in Kuhn's terminology: paradigms). In this connection Pannenberg does not deny the importance of the quest for critical testing and falsification – unlike Kuhn, he still sees falsification as more important than confirmation –, but he argues that it must be done within the proper framework: the encompassing theoretical framework always has to be taken into account, even though, for practical reasons, not every scientist will be able to examine it in detail. Nor does Pannenberg therefore agree with Popper that, whenever an instance of falsification occurs, the theory in question should never be adjusted in order to prevent it from having to be discarded. When we *improve* a theory by making an adjustment to it, there is, after all, nothing wrong with 'saving' it in this way. For clearly, there is no getting around the fact that scientific testing always takes place, not through direct one-to-one falsifications, but by comparing the ability of various theories to do justice to *all* the available material.[6]

In Pannenberg's own conception of a scientific theology, the category of *the whole* ('das Ganze') therefore plays a decisive role. There is every indication here that Pannenberg has fruitfully combined his reading of Kuhn with Hegel's legacy – after all, Hegel also thought strongly in terms of large wholes, ultimately even of history as a whole. However this may be, according to Pannenberg theology is the science concerning God, that is to say, concerning the all-determining reality – for that is, in his view, what is generally meant by the word 'God'. Theology is, then, concerned with the question of how all separate entities and events can be situated meaningfully in the context of their relationship to God (*sub ratione Dei*). Ultimately, this will only become evident once world history has come to an end. It is, after all, only then that the comprehensive framework of our reality, against which all our theories should be tested, will be completed. Consequently, all our present theological statements are pro-visional. In theology, starting from various partial experiences, we reach out in anticipation towards the ultimate state of affairs – and it is this ultimate state of affairs which will, in the end, decisively verify or falsify our theological theories (here, Pannenberg takes over John Hick's view concerning eschatological verification).

The most important partial experience from which Christian theology creates a scheme of where we stand with regard to our reality is the resurrection of Jesus Christ as reported in the Gospels. In these reports (which Pannenberg regards as

6 *Theology and Philosophy of Science*, 56f., 66. In fact, Pannenberg implicitly takes up here the Duhem-Quine thesis; there does not appear to be any influence from Lakatos, who also emphasizes this point.

broadly historically reliable), we see the extent to which the God whom Jesus Christ called his Father is indeed the all-determining reality. The shift into a new mode of existence, which took place on a small scale in Jesus' resurrection, will in the course of time take place on a large scale, thereby revealing the meaning of history in its totality. This, at least, is the hypothesis that Christian theology defends when it comes to the question of how, ultimately, all the pieces of the puzzle called 'reality' or 'history' will fit together.

As we have said before, it is not our concern here to go into the details of Pannenberg's conception of theology. At this point, we do, however, come across an important methodological limitation in that conception. This limitation is closely connected with the fact that, at a crucial juncture, Pannenberg has not distanced himself sufficiently from the attractions of critical rationalism. And he can hardly be blamed for this. On the contrary, the way in which Pannenberg, together with some of his countrymen,[7] managed at the beginning of the 1970s to relate developments in the philosophy of science of the time to theology, and were even able to make a separate contribution to the philosophy of science, can only be called impressive. Now that these developments have gone further, however, we cannot but note, in retrospect, how Pannenberg (in spite of himself?) remained trapped at a number of points, in positivist ideals of science that have, in the meantime, turned out to be untenable.

The point at which this can be most clearly seen, because it leads to an inherent limitation in his theological method, concerns his resistance to any subjective *commitment* in theology. Pannenberg has never ceased to be deeply impressed by the criticism of Popper and some of his students of the ease with which many who theologize have recourse to a personal source of inspiration immune from further interrogation. Whether it be the Bible, the Christian doctrinal tradition or some other authority, Pannenberg regards it as plainly unscientific and harmful to the cause of theology to appeal to any such source of inspiration without being prepared to have it questioned by level-headed and detached scientific criticism. According to Pannenberg, Christians, and Christian theologians in particular, should never simply presume the truth of their faith.[8] Their claims have the status of hypotheses which, like (other) scientific hypotheses, have yet to be definitively confirmed or refuted. No appeal to an unassailable authority can hide the fact that the truth of their beliefs or theological claims is as yet undecided. Thus Pannenberg

7 The work of Gerhard Sauter in particular deserves mention. Cf. the collections of essays he edited, *Theologie als Wissenschaft*, Munich 1971; and *Wissenschaftstheoretische Kritik der Theologie. Die Theologie und die neuere wissenschaftstheoretische Diskussion*, Munich 1973. For a discussion of Sauter's contributions, see Wentzel van Huyssteen, *Theology and the Justification of Faith*, Grand Rapids 1989, 101-121.

8 Cf. also Pannenberg, *Systematic Theology* 1, 48-61.

is also sharply critical of the way in which Karl Barth took (what he saw as) God's revelation in Jesus Christ as the incontestable starting point of Christian theology.[9]

One can, of course, understand Pannenberg's concern in criticising what he calls 'dogmatic theology'. He is concerned lest theology become isolated from what is generally accepted, by withdrawing into a bombproof bastion of fideistic starting-points about which no further rational discussion is allowed, and to which the rule of 'take it or leave it' applies. Pannenberg's fear is that this would lead to an isolated theology that condemns itself to a barren ghetto-mentality. Pannenberg, who did not grow up as a believer himself, wants Christians (or, at least, the theologians among them) to be capable as far as possible, in the spirit of 1 Peter 3:15, of explaining to others why they believe as they do, and with what degree of certainty. Thus it is precisely Pannenberg's missionary attitude that leads him to adopt such a critical attitude towards any 'revelational positivism'.

For the same reason, Pannenberg wants to see God, not in the first place as a *datum* of theology, but rather as a *dandum*, not as a fixed given, but as always something yet to be given or, in his own words: not as a fixed certainty, but a problem.[10] Those who want to define theology as speaking of God dogmatically rather than problematically remain irrevocably trapped in a kind of fideistic subjectivism in which outsiders would never show any interest. By contrast, speaking of God as a problem (does God exist or not, and if so, how and where?) is something to which, today, just about anyone can relate. In doing their work, theologians should therefore strive to leave any faith-commitment they may entertain out of consideration, and view God simply as a hypothesis. Although God may well play a constitutive role in their personal lives, and ultimately even as a motivation for their work, they should never appeal, in their theological reflections, to their personal, faith-based conviction. In technical terms: Their subjective commitment can play a role in the heuristic context of discovery, but not in the probative context of justification.[11]

However much sympathy we may have with Pannenberg's motives, it nevertheless seems that his views concerning the nature of faith and theology are misguided at this point. More specifically, they fail in two respects, namely, from the perspective of faith and theology, and from the perspective of the philosophy of science.

9 Cf. e.g. Pannenberg, *Theology and Philosophy of Science*, 265-276 (cf. above, section 5.2).
10 '... that God is the object of theology only as a problem, not as established fact'; *Theology and Philosophy of Science*, 300.
11 *Theology and Philosophy of Science*, 321.

First of all, it goes against the nature of religious faith to regard the existence of God as a hypothesis that may or may not be confirmed, or turn out to be true, in the future. For the believer, God is, after all, not a problem, but a reality, and even, according to Pannenberg's own definition, the most important, all-determining reality. Therefore believers address themselves to God in special types of language like worship, praise, prayer, and expressions of unqualified trust. Theologians who are also believers cannot abstract from the situation in which they find themselves. For them, faith is simply not an open hypothesis, and it is likewise impossible, therefore, to pretend that it is. It is part of the unique character of faith that its certainty does not depend on contingent historical arguments or other scientific probabilities, but is of a wholly different nature.

In this sense, Christian faith rests, according to Wilfred Joest, on the conviction that God has expressed himself decisively in Jesus Christ. Through God's action in Jesus Christ, believers know themselves to be justified, but also to be called to discipleship and guided by certain norms in their everyday life.[12] All in all, this basic presupposition has so many consequences for their life that they could not possibly regard it as merely hypothetical. It is possible, of course, to derive all sorts of theological hypotheses from it, and 'release' these for scientific investigation; but as far as the basic presupposition as such is concerned, it is only possible either to let oneself be completely determined by it, or not at all. To let oneself be determined by it with a proviso is out of the question.[13] Thus it also goes against the nature of faith to demand of theology that it does just that. Even though, as we have pointed out, a distinction needs to be made between faith as an existential matter and theology as a reflexive (or rational) one, the difference between them cannot be taken so far. For, to put it sharply: only non-believers could then be theologians. A believer can simply not speak in a hypothetical sense about the most determinative and reliable reality in his or her life. And, no matter how different theology may be from faith, this does not, of course, change when the believer in question also happens to be a theologian.

12 W. Joest, *Fundamentaltheologie. Theologische Grundlagen- und Methodenprobleme*, Stuttgart 1974, 253.

13 'Man kann sich darauf unbedingt oder gar nicht, auf keinen Fall aber mit hypothetischem Vorbehalt einlassen' ('It can be adopted unconditionally or not at all, but in no case with hypothetical reservation'), *ibid.* Cf. for Pannenberg's response his *Systematic Theology* 1, 57 n.127, where he argues that the act of faith is unconditional indeed, but that still the statements of faith are hypothetical. It is unconvincing, however, to disconnect the *fides qua* from the *fides quae* in this way, because the act of faith is always directed at an object, of which at least some things are believed to be true, e.g. that it exists, that it is reliable etc. For Pannenberg's (as far as I can see unaltered) later view on the issue, see his *The Historicity of Nature*, ed. by Niels Henrik Gregersen, West Conshohocken 2008, 8.

It might be objected that theology thereby departs from the accepted understanding of science. Yet that is highly questionable, for the other sciences too accept the existence of their object of study without, on the one hand, comprehending it totally or, on the other, regarding it as hypothetical. The biologist, for example, does not grasp exactly what 'life' is, yet in his research makes that very phenomenon his point of departure. If a disanalogy were nevertheless at stake here, then one would have to say, with Barth and others, that nothing can be done about it, since it just so happens that theology has a unique object, and is therefore also a unique science (*sui generis*). After all, what could be more scientific than to attune the methods of study to the special character of the object studied?[14] For theology, this means that it takes its point of departure in God and God's revelation, and does not regard these as hypotheses which need to be proven. This would in fact contradict the priority of God's action[15] and, in the final analysis, even God's divinity. God is, after all, primarily an addressing subject, and not an object to be investigated.

D.Z. Phillips writes somewhere in a somewhat ironic tone that the poet of Psalm 139 would have been well advised to express himself more accurately if we indeed had to view faith in God as a hypothesis which required further proof. For instance as follows: 'Whither shall I go from thy Spirit? Or whither shall I flee from thy presence? If I ascend up into heaven, it is highly probable that thou art there: if I make my bed in hell, behold it is cumulatively apparent that thou art there also'.[16] In short, when we view religious faith (for, presumably, this observation applies not only to Christian and Jewish faith) as a scientific hypothesis in need of further testing, we distort the unique character of that faith.

Secondly, Pannenberg's view concerning the hypothetical nature of faith and theology is not only deficient from this vantagepoint, but also from the perspective of the philosophy of science. Although he appears to share Kuhn's view that our scientific theories are always embedded in a particular paradigm – which as such can never be wholly accounted for in a neutral, paradigm-independent way, and which is always 'prejudiced' in one way or another – he nevertheless demands of theologians that they suspend their personal faith-commitment when

14 See Karl Barth, *The Doctrine of the Word of God* (= *Church Dogmatics* I 1), 3-11 on theology's 'Sachgemässheit' – which should not be translated as 'objectivity' (7) but as 'fitness-for-the-object' – and compare our discussion of Barth's views in section 5.2 above. Along similar Barthian lines also T.F. Torrance, *Theological Science*, London 1971.

15 Cf. for further elaboration of this point with a view to the doctrine of revelation, Ronald F. Thiemann, *Revelation and Theology. The Gospel as Narrated Promise*, Notre Dame 1985.

16 D.Z. Phillips, *Faith after Foundationalism*, London 1988, 10.

embarking on their work. Yet, in doing so, Pannenberg remains trapped in the myth that an objective starting-point may be possible, or, to put it more precisely: that we could succeed in adopting such a starting-point as soon as we take leave of the actual context of discovery, and enter the theoretical context of justification. Yet we have already seen (in the conclusion to section 2.3) the extent to which the newer philosophy of science has discredited this traditional distinction between the contexts of discovery and justification. We found this confirmed, moreover, in Pannenberg's own work, when he pointed out, contra Popper, that the criteria with which we test scientific hypotheses are never totally objective, but can only rest on intersubjective agreement. The same point can be made by saying that these criteria rest on a shared *commitment* on the part of the group of scientists in question. Such a commitment certainly need not by definition be irrational.

Therefore, despite himself Pannenberg has failed to wrestle free from the unjustified (because unattainable!) demands made on scientific practice in a positivistic climate – a climate which continued to be pervasive in critical rationalism. No wonder, then, that J. Wentzel van Huyssteen speaks, in this connection, of the most problematic element in Pannenberg's theory of knowledge.[17] Van Huyssteen also succeeds, moreover, in locating the source of this confusion in Pannenberg's work very accurately: Pannenberg has been all too impressed by critical rationalist William Bartley's criticism of the irrational way in which theologians tend to retreat back to their personal religious convictions. Bartley regarded this as unscientific, and advocated, over against such a strategy, a so-called 'commitment to non-commitment'.[18] Following Bartley and others, Pannenberg regarded the danger of a self-imposed isolation of theology as so serious that he not only worked out, in great detail, how theological hypotheses ought to be testable (i.e. with reference to the condition that they should make maximal sense of our experience of reality), but also required the theologian, in evaluating these hypotheses, to discount his or her own faith-commitment. Thus, he supported Bartley's plea for a 'commitment to non-commitment'. This plea is, however, in conflict with the fact that *every* set of criteria by means of which one may wish to test hypotheses, both in science and in theology, is unavoidably determined by intersubjectivity and ultimately even by some view of life. As we have seen, none

17 J. Wentzel van Huyssteen, *Essays in Postfoundationalist Theology*, Grand Rapids 1997, 66 ('... the most problematic element of Pannenberg's epistemology'); cf. also idem, *Theology and the Justification of Faith*, 89.

18 Van Huyssteen, *Essays in Postfoundationalist Theology*, 58, 64-69. Cf. William W. Bartley, *The Retreat to Commitment*, London 1964. Bartley's influence on Pannenberg is strengthened by the theology-critique of another Popperian, Hans Albert; see his *Treatise on Critical Reason*, Princeton 1985, esp. 132-164 (original German edition: Tübingen 1968, 104-130).

other than Pannenberg himself has emphasized this in his book (taking his cue from Kuhn's paradigm theory).[19] We cannot simply lift ourselves out of the paradigm within which we operate in order to adopt a neutral point of view; it is already difficult enough for us to gain awareness, even to a limited degree, of the paradigmatic determinedness of our thinking. Yet that is precisely what Pannenberg seems to demand, without showing how it could be done.

Of course, the way in which the prevailing approach to forming theories in science is determined by given paradigms is quite different from the way in which someone's theology is determined by his private choice of faith. However, the point at issue, as Van Huyssteen correctly points out, is that the role played by basic personal convictions or commitments cannot simply be excluded, not even in the context of justification. 'The personal involvement of the scientist, in the paradigm out of which he or she lives and works, always plays a role, not only in the so-called context of discovery, but also in the context of justification'.[20] Possessing such fundamental (inter-)subjective basic convictions, the truth of which cannot be demonstrated in any simple sense, cannot therefore be regarded as a scientific vice. What one can do, though, is to become as aware as possible of the role played by these fundamental convictions, and to expose the more tangible religious viewpoints and statements in which they become manifest to critical argument.[21] It is precisely at this point that the scientific attitude can be distinguished from the irrational attempt to immunise oneself against criticism. Yet Pannenberg does not succeed in throwing light on the role of subjective basic convictions in science and theology, since he approaches it in an inconsistent manner, in part using critical-rationalist parameters, and then again from the perspective of Kuhn's newer philosophy of science.

We conclude, then, that Pannenberg had, already at an early stage, discerned the great significance of the newer philosophy of science, and managed, from that perspective, to point to more possibilities for a scientifically responsible discourse on God than had been the case for a long time before him. Yet, at the same time, his insight into the significance of the newer philosophy of science could not have a great effect because, at a crucial point, Pannenberg did not manage to extricate himself fully from the positivist legacy of critical rationalism. Therefore, we shall now consider a second recent conception of a scientific theology, namely that of Nancey Murphy, which has already been mentioned briefly in the previous chapter. More so than Pannenberg, Murphy takes as her point of departure

19 Pannenberg, *Theology and Philosophy of Science*, 55 (cf. above, n.5).
20 Van Huyssteen, *Essays in Postfoundationalist Theology*, 59.
21 Van Huyssteen, *Essays in Postfoundationalist Theology*, 68. A good example of this has, incidentally, been considered above, in Joest, *Fundamentaltheologie*, 253.

the newer philosophy of science, in her case especially the version put forward by Imre Lakatos. From that perspective, she is able to indicate more clearly than Pannenberg the role that may or may not be played by subjective and intersubjective basic convictions in theological and (other types of) scientific inquiry.

6.3 Theology as science in the work of Nancey Murphy

In her survey of the recent history of the philosophy of science, Nancey Murphy also comes to the conclusion that the prospects for classical theology, which initially seemed so bleak, have gradually and decisively improved. As the philosophy of science increasingly took account of the actual practice of scientific research, new possibilities emerged for positive analogies and relationships between science and theology.[22] Murphy even dares, following on from this, to assert that, from a methodological perspective, theology can (potentially at least) no longer be distinguished from the natural sciences (198). After all, according to Murphy, theologians too work – unfortunately without on the whole being aware of it – within the type of research programme, the ingredients and internal relations of which Lakatos has so carefully outlined (183).

The reasons why theologians sometimes eventually (have to) abandon their programmes also have parallels to those which Lakatos pointed to for natural science programmes. As an example, Murphy discusses the theological programme of the so-called Roman Catholic modernists (Alfred Loisy, George Tyrrell and others) of the early twentieth century (88-129). In the long run, its proponents no longer managed to discover new, independent data ('novel facts') that could confirm the theoretical structure (175). By contrast, Pannenberg's theological programme seems to meet Lakatos' criteria for scientific status more successfully (176-178).

What, then, are the characteristics of a theological programme in which satisfaction of the demands for being scientific is consciously pursued? Like a Lakatosian research programme, such a programme has a hard theoretical core which consists of its most fundamental assumptions. These often have a long record in the field of research in question, and have, for generations, been confirmed by countless observations. The core thus also has a great capacity to inspire: those who are convinced of its truth strive to show that all sorts of things can be best understood when illuminated from the perspective of this core. The hard core therefore operates as a kind of organizing centre: those executing the research programme find the hard core so worth defending that they construct a protective fence of auxiliary hypotheses around it. Phrased negatively, one could say that these are meant to shield the hard core from

22 Murphy, *Theology in a Scientific Age*, 85; in this section, numbers between brackets in the main text will henceforth refer to pages in this book.

various forms of criticism (the so-called 'negative heuristic'). This is necessary, for 'without the decision to pursue and defend some central theory, there is no program at all' (60). Positively, however, by means of such flexible auxiliary hypotheses it is attempted to keep intact the link between the core and 'hard' empirical reality by adjusting for emerging anomalies (the 'positive heuristic').

We have already seen in chapter 2 how Newtonian mechanics and gravitational laws are an example of a hard core in a comprehensive natural science research programme. Likewise, Murphy argues that the existence of God can constitute an important part of the hard core of a comprehensive theological research programme, together with, say, God's holiness, God's revelation in Jesus, and possibly also the triune nature of God (184). So, in Pannenberg's programme, the 'hard core' is that the God of Jesus Christ is the all-determining reality (176). The idea then is that the true significance and meaning of all things and events can only be understood in relation to this core. As the poet in Psalm 36 puts it: 'In your light, we see light' (NRSV).

Clearly, certain questions immediately arise here. Can we let a religious belief take the same place as that occupied in the original Lakatosian programme by deeply entrenched scientific convictions? Van Huyssteen for one is of the opinion that Murphy already takes a wrong turn at this point. According to him, locating theistic assumptions in the hard core of a theological programme does not yet guarantee a scientific theology.[23] Rather, it seems that Murphy retreats here into a subtle form of the very foundationalism she rejects, by taking her point of departure in ultimate, unprovable certainties for which, allegedly, no reasons have to be put forward. Van Huyssteen regards such a position as irresponsible and as one which leads ultimately to epistemological relativism. He even suspects Murphy of 'crypto-fideism' in this connection.[24] Yet, for three reasons, this criticism is misplaced.

First of all, it is certainly true that the manner in which Murphy gives content to the hard core of her theological programme does not guarantee its scientific nature. Yet that is not what she claims. For her, in accordance with Lakatos' methodology, scientific status is determined one way or the other by quite different factors, especially by the progressive nature of the programme. The break between the newer philosophy of science (among which the Lakatosian research programme theory) and the older schools (especially logical positivism) involves precisely the fact that

23 Van Huyssteen, *Essays in Postfoundationalist Theology*, 84. 'Using the methodological framework of Lakatos' model to locate theist assumptions ... in the unassailable hard core of a theological program certainly does not guarantee probable [i.e. scientific, GvdB] reasoning'.

24 Van Huyssteen, *Essays*, 88, 89f. Broadly speaking, a 'fideist' is a believer for whom faith (Latin: *fides*) is, and may be, thoroughly irrational.

the scientific nature of various isolated statements cannot be determined apriori, but only via a difficult and often long drawn-out process of testing, in which the broader framework that encompasses the statements in question is very decisive.

Secondly, Van Huyssteen rejects theistic assumptions as good candidates for the hard core of a theological programme without indicating the criteria on which basis he does so. Interestingly, Murphy herself fails to a large extent to give an account of this either. What is clear, in any case, is that it does not make sense, from a Lakatosian perspective, to demand that the correctness of the assumptions that are admitted into the hard core have to be *proved*. Whether these assumptions are correct is, after all, precisely what is at issue in the research programme. If the correctness of the hard core were already evident to everyone to begin with, then presumably executing the research programme in question would only have limited value. For the same reason, it is wrong to demand that the assumptions that together constitute the hard core at least be 'generally' accepted or generally acceptable. After all, Lakatos assumes (more so than even Kuhn does) a situation of normal science, where two or more alternative research programmes compete with one another, with each having its own adherents. We cannot therefore deny theistic assumptions access to the hard core of a theological programme on the grounds that they are unproven or not generally accepted.

Thirdly, it is highly questionable whether it is at all possible to formulate criteria which could be used to determine what sorts of assumptions should or should not be allowed into the hard core of a research programme. For such a view presupposes that this selection involves a conscious process. Yet, in reality, it involves (at least in *comprehensive* research programmes) ideas that are so deeply rooted in the researchers' experience of reality that they remain partly unconscious. And, in so far as it is possible to be conscious of the assumptions in question, they seem so self-evident that it one cannot be said to *choose* whether or not to make them part of the hard core of the programme. Researchers will simply find themselves in a situation where these assumptions are regarded as self-evident and incontestable. For the most part, it is a question of simply having grown up with them, so that the hard core seems to be confirmed countless times in a variety of ways (cf. the hard core consisting of the Newtonian theories which we have discussed already).

Accordingly, scientists who work within an evolutionist research programme do not constantly aim to expose the assumptions of the theory of evolution to criticism, attempts at falsification or other tests.[25] They are much too personally committed to these assumptions to do this. At the same time, however, it is true that they are faced with various as yet unresolved issues, among which a number of anomalies that are in tension with the hard core. It is precisely this fact that inspires

25 I owe this example to Stenmark, *Rationality in Science, Religion, and Everyday Life*, 120.

the execution of the research programme. It is therefore not fair to blame someone at the very start for including the wrong assumptions in the hard core. After all, the fact that they are wrong can only be seen from the outcomes of the programme in question. The same rule applies here as everywhere else: 'the proof of the pudding is in the eating'. What the hard core of a comprehensive research programme looks like is to some extent an existential question. It is not without reason that, according to Lakatos, hard cores often have a metaphysical background (199).

So, in this regard, Lakatos and Murphy have clearly gone beyond Pannenberg: The fact that scientific and theological programmes can ultimately be traced back to (inter)subjective commitments should not be denied or regarded as a disqualification, but rather should be taken seriously by accounting for it.[26] And this is precisely what happens in the research programme inspired by them. Here, it must be made clear to what extent the organising centre is better able than its alternatives to do justice to the diversity and complexity of reality. Given all this, it is entirely reasonable that fundamental tenets of faith such as the existence of God or the love of God may form part of the hard core of a theological research programme. After all, for those involved, these faith-convictions have precisely the existential self-evidence that befits a hard core, while, at the same time, they also recognize that this self-evidence is not so obvious to everyone, to say the least, and that the necessary anomalies in fact also exist. This faith therefore never ceases to search for understanding, in accordance with the old motto of medieval theology (*fides quaerens intellectum*).

How should we then proceed, according to Murphy, in order to make progress in this regard? First of all, we shall have to make continual additions, replacements or modifications in the surrounding belt of auxiliary hypotheses, in order to deal adequately with the anomalies that arise. Ideally, this handling of possible falsifications will not occur *ad hoc*, but rather on the basis of certain rules that safeguard the hard core from modification: together, these rules constitute the so-called 'negative heuristic'. In this regard, Murphy sees church doctrine as such a series of guidelines that indicate the direction and limits of theological creativity (185). For instance, if we regard the presence of suffering and evil in the world as an important anomaly for belief in God, then surely we should find some form of theodicy in the protective belt, that is to say, a 'theory' that indicates how belief in God's omnipotence and goodness need not be given up despite all the evil in the world.

26 It is, then, remarkable that, on the one hand, Van Huyssteen criticises Pannenberg for not allowing a legitimate place for this type of basic assumption (Pannenberg 'fails to confront the vital question of the intrinsic role of the theologian's subjectivity', 64), yet on the other, a few pages later, criticises Murphy for doing just that: 'This leads one to suspect ... a deeper and prior commitment' (83). Should this be attributed to the fact that Van Huyssteen's book consists of previously published essays, the time span between the first and the second essay quoted here being seven years?

However, it is not only necessary to deal in a more or less consistent fashion with falsifying experiences (that happens just as effectively in, say, Marxism), but, secondly, also to *broaden* the theological programme's grasp of empirical reality. This requires plans and guidelines regarding possible new avenues of explorations and elaborations of the programme; together, such plans and guidelines constitute the programme's 'positive heuristic'. It is, after all, only in this way that it can become more evident that the hard core of the programme (e.g. the love of God) does in fact point to something real, and can continue to establish relationships with empirical reality. As we have seen, Lakatos thinks that the empirical content of a programme is especially increased when, by using a particular set of auxiliary hypotheses related to the core (which themselves often constitute a limited research programme), there is success in predicting a novel fact and, if possible, also in discovering it. This is a rather strict criterion – novel facts do not arise all that often in science, and can even remain undiscovered for considerable periods of time – which is nevertheless important if a programme is to remain vital and fully scientific. Can such 'new facts' (which, as we saw in chapter 2, may also include old facts cast in a new light) also appear in theology? Murphy rightly believes that they can.

Let me give an example that can make this clear. If we take as our point of departure a hard core that includes the existence and love of the God of Israel, then a possible elaboration might lie in the *faithfulness* of this God to what He has initiated. From that perspective, one might venture to predict that the moment will never arrive when there will be no Jews living on earth any longer, or that the people of Israel will never disappear completely from history. Many Jewish and Christian theologians have indeed made this prediction (the latter especially since the so-called Further Reformation in the eighteenth century). However, by the end of the Second World War, when the horrors of the holocaust were forcefully brought home, the credibility of this prediction must have sunk to an all-time low. Yet when, three years later, in 1948, the state of Israel was established in Palestine, and emerged with increasing strength from a spate of successive attacks from neighbouring countries, it constituted a sublime example of a *novel fact*: a shocking, totally unexpected event, which just happened to constitute an impressive confirmation for any theological programme that took as its point of departure the faithfulness of the God of Israel to his people.[27]

For her part, Murphy identifies yet another way in which the empirical quality of a theological programme can be tested, and possibly increased, in a way that is

27 In fact, as far as I know, the emergence of the state of Israel was not predicted, but it was interpreted theologically from the start as a public sign of God's faithfulness. See e.g. Paul M. van Buren, *A Christian Theology of the People Israel* (A Theology of the Jewish-Christian Reality, Part 2), New York 1983.

less dependent on a one-off event like the founding of the state Israel in 1948, but which allows for the possibility of repeated testing, as occurs in the natural sciences. She takes as her point of departure the enormously large number of religious experiences which we know of – a number that still increases daily – and then asks the question: How can we distinguish, here, between the authentic and the inauthentic? Or to put it differently: How can we distinguish between experiences that only tell us something about a particular religion, or about the psychic disposition of people, and experiences that say something about who God is? In this sense, she considers religious experiences (along with historical events, Scriptural texts etc.) as important *data*. From a structural point of view, she argues, there is no difference, in this regard, between a natural scientific programme and a theological one. Just as physicists can select from a multitude of sensory experiences the data that are significant for, say, understanding the nature of matter or energy, theologians too ought to select from religious experiences the data against which their theological doctrines can be tested. In this regard, then, theologians should go further than psychologists of religion, who register and analyse religious experiences, but cannot pronounce from within their discipline on their veracity.

At this point, Murphy sides with Christian tradition, which, according to her, possesses a wealth of answers to questions such as how we can discern a word or action of God, and how we can distinguish between experiences that involve a true encounter with God (and that as a result can include authoritative insights concerning life or doctrine), and experiences that have to be explained differently (132). She thinks in this connection of the Christian practice of the so-called 'discernment of spirits', and provides a survey of how this practice[28] takes shape in the work of three influential representatives of different confessional traditions. Thus, she discusses the work of Jonathan Edwards, Ignatius Loyola and Pilgram Marpeck as representatives of, respectively, the Reformed, Roman Catholic and (her own) Anabaptist tradition. Besides, she pays attention to the early church. The survey that emerges from this reveals a rather broad consensus concerning the criteria that are of importance in the practice of 'Christian discernment'. According to Murphy, it is at least clear that the following are among the criteria that religious experiences should satisfy in order to qualify as God-given (152):

(1) agreement with the apostolic witness;
(2) the development of the image of Christ in those who have undergone the experience; this image is especially recognisable as a liberation from sin, and a manifestation of the fruits of the Spirit (love, peace, joy etc.)

28 Murphy employs the concept of a 'practice' in the sense given to it by Alasdair MacIntyre in his *After Virtue*, Notre Dame 1984², 187.

(3) consensus about it within the faith community (which seeks guidance in prayer on the matter).

In situations where such criteria are satisfied, it is possible, according to Murphy, to point to the work of the Holy Spirit with a significant degree of certainty. According to Murphy, the theologian is therefore justified in transposing his so-called observation language into, or complementing it with, the language of participation (i.e. complementing the observer's perspective with that of the participant). It can then, for instance, not only be scientifically justified to say that 'Sarah Edwards was convinced of God's righteousness, and according to her husband, Jonathan Edwards, it was the work of the Holy Spirit', but also: 'God convinced Sarah Edwards of his righteousness' (159). Of course, a statement of this nature always rests on a particular (theistic) *interpretation* of the data, which cannot be derived with logical necessity from the data themselves; yet this does not constitute an adequate counter-argument, for experience-without-interpretation just does not exist. In the natural sciences as well, an appeal to 'pure observation', which would precede any interpretative theory-formation, has proven impossible.[29] What matters is that the (always already interpreted) data be explained in as consistent a manner as possible by the doctrines and theories that constitute the network of auxiliary hypotheses, and that these auxiliary hypotheses in turn can be well understood from the perspective of the hard core. Programmes in both the natural sciences and theology can be tested on this basis (although such tests are seldom without difficulties).

But should we not simply recognize that the natural science data are objective, whilst that of theology are subjective? Murphy also challenges this widely accepted notion (165f.). She points out that, in relation to observational data in the natural sciences, 'objective' means that, under comparable circumstances, and with comparable means, comparable people will succeed in achieving the same observational results. We have already seen that in the newer philosophy of science the notion of 'hard objectivity' in this way has been replaced (not by pure arbitrariness, but) by the notion of intersubjectivity. Scientific data must be public in the sense that others, who find themselves in similar conditions, can confirm them. In the same way, the Christian practice of the discernment of spirits also involves a public process. There is, so to speak, nothing mysterious about it. To be sure, the individual religious experience-plus-interpretation is private (like any sensory observation-plus-interpretation), but the way in which the faith community judges it can be followed by anyone who can place themselves in their shoes, and who employ the same criteria.

29 Cf. above, section 4.2, on the theory-laden nature of observation.

In this sense, according to Murphy, we can even speak of *repeatability* – the 'hard' fact that traditionally distinguishes research in the natural sciences from that of the humanities. After all, we can, in principle, always re-evaluate the procedures by which the faith community arrives at certain conclusions as to their adequacy and correct (or incorrect) outcomes. (The fact that it can prove difficult, in practice, to render the circumstances completely repeatable, and that the results are not always completely unambiguous, but leave certain loose ends, is also true of the natural sciences, so that there is nothing unique about this complication). Is it then also possible for the theological programme as a whole to be confirmed by novel facts that arise from the discernment of spirits? Murphy thinks it is. As an example, she refers to the theologian who believes, on the basis of old, reliable sources (= the deposit of earlier authentic faith-experiences), that justice is best done to the heart of the Christian faith when the Holy Spirit is referred to with female pronouns (i.e. 'She' rather than 'He' etc.). If, *independently of this theory*, this were to become an established practice in various Christian communities, that would constitute, as a 'new fact', an important confirmation of the theory (167f.).[30]

In a comparable manner, the 'rediscovery' of the missionary character of the Christian church in seventeenth-century Protestantism – the Reformation in the sixteenth century still had a blind spot in this regard – was followed and confirmed, in later centuries, by church-sanctioned missions.[31] Even if this were not, strictly speaking, an example of a 'novel fact', for a causal link cannot be ruled out in this case, then one could still speak of a kind of empirical confirmation. After all, if, according to the apostolic witness, God so loved the world that God sent his only Son (John 3:16) – and let us say that this belongs to the 'hard core' of the faith – then it seems obvious that those who consider themselves addressed by God will also let themselves become more involved with the salvation of the world (a theological auxiliary hypothesis that belongs to the positive heuristic). And if this is correct, then, sooner or later, it would again have to become evident from the fact that the faith community indeed lets itself be sent out into the world to witness and serve. The activities of missionaries and others who give actual shape to this on behalf of their community, could then be quite

30 Perhaps, a more realistic contemporary example here would be the 'evangelical turn' that takes place in many churches in the West and worldwide, often independently of each other. Alternatively, such a comprehensive development can also be explained from a sociological point of view, e.g. as a result of the postmodern *Zeitgeist* (but perhaps both interpretations need not be considered as mutually exclusive)

31 See on this process e.g. Jan A. B. Jongeneel, 'The Missiology of Gisbertus Voetius: The First Comprehensive Protestant Theology of Missions', *Calvin Theological Journal* 26 (1991), 47-79.

adequately understood as a set of empirical *data* that confirms the entire 'programme'. Clearly, that does not constitute *proof* of the correctness of faith (all missionaries may be mistaken), but it would constitute a confirmation of the viability and scientific character of the theological programme in which that faith is seen as rationally justified.

But could not all missionary religions claim this? This leads us, finally, to another obvious objection to Murphy's Lakatosian conception of theology: Does the great variety of beliefs and practices not constitute a strong argument against the reliability of the 'discernment of spirits' that she rates so highly? (166). Does it not turn out in practice that Christians disagree even among themselves on just about everything concerning God's nature, God's will and God's relationship to this world? Murphy recognizes this, but does not conclude from it that the practice of communal discernment is unreliable. On the contrary, she infers from it that this practice has not yet been applied with sufficient frequency and determination. In many churches, she claims, there are countless practices and theological ideas that have never been subjected to a communal, prayerful testing of the spirits to see whether they are God-inspired or not. If that were to happen more consistently, she suggests, there would also emerge more unity between believers in their faith-convictions and attitudes.

This latter idea may well attest to some typically American optimism. Nevertheless, in general I find the manner in which Murphy translates Lakatos' theory of scientific research programmes into theological terms very creative and inspiring. She succeeds in convincing the reader that in faith and theology things in fact function (or, at least, can function) as, according to Lakatos, they do in a well-functioning scientific research programme. This is not so much due to the fact that she understands faith as a more or less probable scientific hypothesis, as we saw in Pannenberg, but, on the contrary, because she shows how, in many respects, scientific hypotheses have the character of faith-claims. Murphy indeed succeeds more satisfactorily than Pannenberg in doing justice to the subjective commitment that constitutes the hard core of faith and theology; she shows why that commitment need not be seen as detracting from the scientific status of the theological research of which it is the core.

Moreover, in her conception of scientific theology, Murphy succeeds in maintaining a high degree of continuity with the faith-tradition of Christianity, remarkably enough even with its most spiritual, pietistic strands. Thus she avoids the mistake we encountered with the logical positivists and, in their footsteps, others as well: the idea that the philosophy of science could prescribe what the everyday scientific shop-floor should look like, or (in her case) that theology would be able to determine from above what the faith of the Christian community ought to look like. Just as the more recent philosophy of science takes as its point of departure existing scientific practices, and strives to reconstruct them rationally, theology is primarily concerned with making the practice of faith comprehensible, by doing justice to its nature, content and meaning.

Next, Murphy prevents that her concept of a scientific theology can only be interpreted as an experience-based theology in the sense which is usually attributed to Schleiermacher. That would, after all, not be a choice that speaks to everyone in the theological spectrum. Apart from religious experiences, and and even as a test of them, she also gives a place to Scripture as a source of theological data (cf. 168-172). Finally, Murphy succeeds in persuasively refuting some obvious objections to her theory which we mentioned above (although some of the issues involved need a more thorough-going discussion).

Nevertheless, at the same time there is one serious problem with Murphy's conception, which might well turn out to be the anomaly that may yet cause the whole programme ultimately to fall apart. This concerns the seamless transition she makes from theological research programmes and religious experiences in a very general sense to those of specifically Christian faith communities. The attentive reader will not have missed this transition in the argument we have just described. Remarkably, Murphy offers no justification for this, and simply leaves aside the whole problem of contemporary religious pluralism. Yet, in his critical reaction to Murphy's proposals, Wentzel van Huyssteen has rightly pointed to this problem. As we have seen, Van Huyssteen wrongly reproached Murphy for having fallen prey once again to foundationalist frameworks of thought. Quite deliberately, Murphy opts for the horizontal metaphor of a core and a surrounding belt, rather than the vertical one of a foundation of infallible human knowledge with the solid structure of science built upon it. If anything has become clear from the history of the philosophy of science, it is that human knowledge is, by definition, fallible, and Murphy has clearly grasped that (which is why she does not write that the hard core is unassailable, as Van Huyssteen claims, but only that we should not give it up easily).[32] Van Huyssteen is, however, right when he objects that Murphy leaves no room for real, intersubjective communication about theological claims concerning knowledge, as she derives the criteria for distinguishing between 'authentic' and 'inauthentic' religious experiences and insights entirely from the Christian tradition. In his view, she thereby in fact opts for theological isolation; in the worst case, this leads to a (typically Anabaptist?) ghetto-theology, which serves only to confirm the correctness of one's own point of view.

A theology that seeks a stimulating dialogue with agnosticism and atheism (and, we should add, with the adherents of non-Christian religions) would, according to

32 Van Huyssteen, *Essays in Postfoundationalist Theology*, 86; Cf. for Murphy's theory of knowledge also her article co-written with James McClendon, 'Distinguishing Modern and Postmodern Epistemologies', *Modern Theology* 5 (1989), 191-214.

Van Huyssteen, not limit itself to designing theories that are tested against criteria which are only recognised within one's own faith tradition. After all, it would then risk forfeiting its public function, and, in spite of itself, become isolated from the (other) sciences. Moreover, it would become quite unclear how we should choose between all the theological research programmes competing with one another, something that was quite essential for Lakatos. If everyone were to derive the testing criteria for their own programme from their own tradition, then it would no longer be possible to say what sort of criteria we could use to judge whether one programme is better or more suitable than another. What we should therefore look for, according to Van Huyssteen, is what he calls *transcommunal* criteria for theological research, i.e. criteria that transcend the boundaries of a specific (faith) community, and also provide insight for members of other communities.[33]

Let me give an example of what, in concrete terms, may be at stake here. According to Murphy, belief in the resurrection of Christ would have to be justified from within a consensus in the Christian faith-community concerning the truth of first century utterances of people claiming to have met the risen Christ. Yet someone outside the Christian faith-community would presumably not be particularly impressed by such a consensus; he would, for instance, object that people probably belong to the community in question precisely *because* they believe in the resurrection of Christ (amongst other things). The scientific 'quality' of that belief should, then, not be rated that highly. Suppose, however, that we were to advise this person from outside the Christian community to place the ancient reports concerning the resurrection next to one another, and test them according to accepted historiographical methods. We might then ask him whether the most adequate explanation for these reports is not in fact the actual resurrection of Christ. In that case, we would be appealing to criteria (in this case of a historiographical nature) that transcend the boundaries of a particular faith-community, thereby making real intersubjectivity and a fully scientific debate possible.

It must surely be possible, from time to time, to find such community-transcending criteria, and theology would do well always to look for them (as it has, in fact, constantly been doing since the second century Apologists). It is highly questionable, however, whether, on the basis of such criteria, a *methodology* can be developed that would enable us to judge, independent of any view of life, which of the competing theological programmes and traditions are making

33 Van Huyssteen, *Essays*, 87: Van Huyssteen's exact terminology is 'transcommunal or intersubjective explanations' (following Clayton, *Explanations*, 113ff.), but it seems to me that this transcommunality should concern in particular the *criteria* by which explanations are to be tested (while the explanations or theories themselves may well emerge from one particular community).

progress, and which, by contrast, are degenerating. Or perhaps this is not a real question at all, since we should rather frankly admit that it is clearly not the case. For if we tried to develop such a methodology, it would soon turn out that situations, experiences, practices et cetera can always be evaluated in different ways, and that it is not possible to achieve intersubjective consensus about them.

In chapter 2, we already saw that Lakatos' theory of scientific programmes is ultimately too optimistic: even a degenerating programme can turn out, after a long time, to be progressive, and to deserve being rated much more highly than its competitors. If we apply this to theology, which, for better or for worse, concerns itself with 'ultimate questions', we should recognise that it is even more difficult to arrive at definitive judgements here. For instance, we cannot simply claim, on the basis of sociological data such as secularisation in Western Europe, that the Christian theological research programme has definitively proved inadequate or inferior to competing humanistic or scientistic programmes.[34] For clearly, truth is not something that can be established by a majority vote, and, since criteria deriving from views of life always play a role, consciously or unconsciously, in judgements of this type, we will continue to evaluate the empirical value (and therefore the persuasiveness) of the various programmes differently in different communities.

We arrive at the conclusion, then, that, neither in the sciences, nor in theology rational progress by means of the succesion of research programmes is guaranteed. A linear development by which we are gradually shown more clearly what reality is really like cannot be secured in either case (in theology – due to the greater complexity of the questions it deals with – even less so than in the other sciences). The notion of scientific status should therefore be separated from this modern idea of progress. But how much of it still remains then? Are we not, in that case, at the mercy of the complete arbitrariness of thoroughly historically and contextually determined insights that decisively shape our particular traditions and communities? From a logical point of view, however, this conclusion does not follow. In the final section of this chapter, we will try to find out – now from a Kuhnian rather than a Lakatosian perspective – how we can indeed have 'transcommunal', meaningful and rational *discussions* about what does or does not count as properly scientific work in theology. Even though a prescriptive methodology, with an elaborate set of rules and criteria arrived at apriori,[35] always turns out to clash with an intractable

34 For a more extensive argument on this point, see Gijsbert van den Brink, 'Cultural Plausibility and Christian Theism', in: H.A. Krop et al. (eds.), *Post-Theism. Reframing the Judaeo-Christian Tradition*, Leuven 2000, 265-275.

35 In this connection, Stenmark, *Rationality*, 39ff., speaks of 'the formal view of science', an approach that he also associates with Lakatos, and that he deems inadequate. Ultimately, Stenmark draws more inspiration from Kuhn (381), and argues that Kuhn's approach need not lead to relativism.

reality that does not allow itself to be easily grasped, this does not yet mean that we must automatically end up in a relativistic philosophy of science.

6.4 In Kuhn's footsteps: Christian faith as a paradigm

In accordance with our findings in chapter 2, we shall, in our sketch of theology as a scientific discipline, take our point of departure even more clearly than either Pannenberg or Murphy did in the awareness that we live in a post-Kuhnian era. This means that we shall leave behind the old dream of a formal determination of science by means of a set of fixed and immutable methodological criteria that should equally apply to all disciplines. There simply is no such set of criteria, and there will never be one (if only because of that word 'immutable'; it is practically unimaginable that, whilst our scientific knowledge is continually on the move, our metascientific insight into what count as good criteria should remain completely static). That is not to say that we can no longer have anything at all to say on the question of whether a particular intellectual practice does, or does not, satisfy scientific standards. In this connection, we continue to resist the relativistic interpretation of Kuhn's legacy. It is not without reason that, apart from scientific revolutions, Kuhn also distinguished long periods of normal science, when people rightly follow all sorts of accepted rules.[36]

Now various authors have already linked Kuhn's ideas to faith and theology. For example, Ian Barbour argued that Christian faith can be seen as a paradigm very much comparable to the scientific paradigms described by Kuhn. Hans Küng interpreted the equivalent of Kuhn's concept of a paradigm somewhat more narrowly, and described the theological conceptions of Augustine, Thomas Aquinas, Luther and Calvin as new paradigms in the history of Christian theology.[37]

In addition to these, yet further links between science and religion, which were partly though not exclusively inspired by Kuhn, have been suggested. Thus, Nicholas Wolterstorff has developed an insight that is characteristic of more

36 In this sense, we can agree with Steve Fuller, when, in his *Kuhn vs. Popper* he qualifies the prevalent image of Kuhn in a conservative direction. However, Fuller's portrayal of Popper as the unrecognised hero, and Kuhn as a sort of second rate author who has 'fallen upwards', is less persuasive given the problems in the philosophy of science (discussed in chapter 2) that Popper failed, but Kuhn succeeded, in solving. In his depiction of Popper as a progressive critic, and Kuhn as patron of the establishment (e.g. on pp. 54-57), Fuller mistakenly ignores the enormously critical impact of Kuhn's disconcerting notions of paradigm shifts and scientific revolutions.

37 Ian Barbour, *Myths, Models, Paradigms*, New York 1974; Hans Küng & David Tracy (eds.), *Paradigm Change in Theology*, Edinburgh 1989, 3-33, 212-219, 439-452; cf. also the study by Basil Mitchell mentioned in footnote 1.

recent philosophy of science as a whole, namely, the *underdetermination* of scientific theories by observational data (which we discussed earlier in connection with the Duhem-Quine thesis), in the direction of a concept of Christian scientific practice. According to Wolterstorff, when there are indeed several theories that accord with the same observations, then Christian scientists should give preference to developing those theories that best fit their view of life.[38] And in the beginning of the present century, Alister McGrath has made the most extensive contribution yet to a concept of scientific theology based on the newer philosophy of science.[39]

In this section, however, we shall not try to subsume these diverse attempts under a typology, or to evaluate them successively as to their merits. Rather, we shall try – drawing lessons from the turn from a prescriptive to a descriptive philosophy of science – to begin by considering the prevailing way of doing theology, in order then to make tentative comments on the conditions under which this can be understood as a scientific practice. I shall thus provide a broad sketch indicating a certain direction, rather than a fully-developed proposal.

If we follow Kuhn in seeing scientific research above all as an attempt to solve all sorts of puzzles by interpreting them in such a way that they obtain a place which provides us with insight within a comprehensive framework, then what kind of 'puzzles' is theology concerned with? Let us list a number of such puzzles or problems as these have been addressed in the theological research of the past and present, and then make a few comments about them.

1. Does traditional human discourse about God refer to a God who really exists, and if so, (how) can this God be known?
2. Did the universe in which we live come into existence intentionally, or is it merely a random fall of the dice?
3. Does the life of the individual person have a meaning and destiny (and if so, what is it), or not?
4. Is human behaviour determined 100% by 'nature + nurture', or is it (ideally) characterised by a certain degree of freedom?
5. Why is it that we experience so much suffering and evil in our world?

38 Wolterstorff, *Reason within the Bounds of Religion*, 72-80.

39 Alister McGrath, *A Scientific Theology*, 3 vols., Edinburgh 2001-2003. However, although McGrath states that he is in search of a scientific methodology for theology, he writes more on various questions of content, and does not enter into discussions about methodology with relevant authors like Pannenberg, Murphy, Clayton, and others. Cf. Willem B. Drees, 'Review of A Scientific Theology, Volume 1: Nature,' *Ars Disputandi* [http://www.arsdisputandi.org] 2 (2002), 3. McGrath's trilogy is thus complementary rather than directly relevant to our project.

6. Are the Jewish people a people like all others, or do they differ more, structurally, from other peoples than those peoples differ from each other?
7. Was Jesus of Nazareth a normal human being, or was he more than that?
8. Is guilt definitive, or can it be undone?
9. Can believers, when they come together in a religious gathering, experience the Spirit of God, as they claim, or can they only delude themselves into believing that they do so?
10. Is there any point in trying to do good?
11. Is there life beyond death?
12. Do chaos, suffering and evil have the final say in world history?

At first glance, these may perhaps seem, for the most part, typically Christian 'puzzles'. On closer inspection, though, this is debatable. To be sure, we cannot claim that all these questions are universal in the sense that every (thinking) human being should be engaged with them. Most inhabitants of Mongolia, for example, would have difficulty making much sense of questions like 6 and 7, and also possibly of quite a few other questions in the list. I do believe, though, that, at least *in many parts of the world* (and at least in the whole Western world), every person who lives his life reflectively would, or at least should, ask this type of questions. With regard to each of the puzzles summarised here, it can be defended that, regardless of their view of life, thoughtful Westerners should be interested in them. After all, most of them are among the deepest of life's questions. For instance, it would surely register even with those who personally have no affinity with the Christian tradition that a (large) number of well-educated and perfectly well-functioning people in their environment view Jesus of Nazareth not only as something more than a normal human being, but even as God, or as an incarnation or unique revelation of God. One may well find such an answer to question 7 bewildering. Yet, in the normal course of things, one would then be all the more interested in the reasons given for it.

It is in a theological research programme or paradigm that such reasons are investigated and judged on their merits. This is not done by testing them one by one, in positivist fashion, against a (purportedly as yet uninterpreted) 'hard' reality, for this does not exist. Rather, it is done by testing them against the question as to whether they are a coherent part of a larger *whole* within which questions like 1 to 12 are answered in an illuminating way. It could be the case, for instance, that a Christian answer to question 7 has far-reaching consequences for the answers to questions like 1, 3, 11 and 12, so that it becomes clear that all these questions can now suddenly be answered in a tightly coherent way. This coherence could even cover all sorts of other aspects of our human experience. The truth of such a Christian answer to question 7 is thus tested, not so much by asking whether it agrees with general human experiences (or, with the 'facts'), but rather by examining the extent to which, in its light, these experiences can be understood as a coherent and meaningful whole. Of course, the reverse may also prove true, namely, that the answer unnecessarily complicates the whole, so that an Islamic or humanist answer appears more adequate.

It is with questions like these that theologians are professionally engaged in disciplines such as doctrinal theology and ethics. Question 1, for example, is dealt with in theological epistemology (i.e. in the so-called 'prolegomena' of theology – which literally means 'things that have to be said first'), questions 2 and 3 in the doctrine of creation, question 4 in theological anthropology, question 5 in the doctrine of sin, question 6 in the locus on Israel, question 7 in Christology, question 8 in the doctrine of atonement and in the so-called doctrine of justification, and in a slightly different way in ethics too, question 9 in pneumatology, ecclesiology and practical theology, question 10 in the doctrine of sanctification and ethics, and questions 11 and 12 in eschatology. The list of questions is far from complete, and it would be quite easy to complement it with other, no less relevant, questions which are addressed in theology. Yet the above list at least touches on more or less the whole spectrum of classical theology. Countless studies continue to be published in all these areas. These studies tend to take the form of (in the ideal case very carefully and intricately worked out) *hypotheses*. Moreover, the more comprehensive studies do not limit themselves to a particular question, but also try to account for the implications for, and links with, other parts of the theological network. To mention one rather arbitrary example from the work the most productive and creative present-day Dutch theologian: Abraham van de Beek's monograph on 'Israel as the people of the suffering Lord' is a very detailed hypothesis concerning the special place of the Jewish people in world history. The answer Van de Beek gives in this book to question 6 turns out to be closely related to his answer to question 7 as elaborated in a previous study.[40]

Naturally, such theological hypotheses turn out not to convince everyone; this will be partly due to the fact that they are always linked to a subjective commitment within a given view of life. Sometimes, that commitment is explicitly present (e.g. when theologians serve a particular denomination), and sometimes it remains more implicit – but it is always there. We have seen, however, that, when it comes to the 'hard core' of research programmes in the natural sciences,[41] we

40 Cf. A. van de Beek, *De kring om de Messias. Israël als volk van de lijdende Heer*, Zoetermeer 2002. Van de Beek's earlier work on Christology has been translated into English as: *Jesus Kyrios. Christology as Heart of Theology*, Zoetermeer 2002.

41 By the way, in social-scientific programmes too, but for the sake of argument I shall concentrate on the natural sciences, which are often regarded as the 'hardest' ones on the one hand and on the much 'softer' theology on the other.

also observe such a commitment. Natural scientists are also, as a rule, very passionate about certain fundamental tenets of which they are convinced, despite the fact that their correctness cannot possibly be shown apriori. The same goes for the adequacy of particular methodological skills that they have acquired on the basis of existing traditions, authorities and even 'dogmas' in their discipline. To demand of them to verify and confirm personally all these basic axioms, principles, skills et cetera, would be to confront them with an impossible task, not only pragmatically, but also in principle, as the history of logical positivism has shown. A much better approach would be for scientists to set about their work with a view to showing how much can be explained and clarified by means of their assumptions concerning methodology and content – which is precisely what they in fact tend to do. The justification of their assumptions initially takes place, so to speak, through their *faith*, and is only confirmed afterwards when they turn out to *work* in reality.[42]

The parallel with theology is already evident here. For clearly, as is well-known, things work similarly there in all these respects. There too, traditions, authorities, dogmas and personal commitments play a significant role. But there too, the rule applies: the proof of the pudding is in the eating. Theological theories also touch, to a greater or lesser extent, on empirical reality, do more or less justice to the whole of experienced reality, exhibit a greater or lesser degree of internal consistency, of coherence with other things we hold true, of intuitive appeal and explanatory power. We may even ask: What, then, are the remaining differences between (natural) scientific and theological discourse? Should we not be wary of overemphasising the similarities to such a degree that we undervalue, or even lose complete sight of, the differences, which, after all, everyone feels are very real? This is indeed a risk we should take care to avoid. Let us therefore try to determine these differences.

It is often stated that a crucial difference between paradigms in the natural sciences and theology respectively lies in the *nature* of the puzzles to which solutions are sought. It may be thought, for instance, that the sciences focus exclusively on questions of fact, whilst theology, by contrast, focuses on questions

42 I have borrowed this phraseology from Lesslie Newbigin, *The Gospel in a Pluralist Society*, 46. Inspired by Michael Polanyi, whom we mentioned earlier, Newbigin (most extensively in this study, 1-65) has indicated persuasively the many similarities between the habitus of faith and that of science. Cf. perceptive remarks such as: 'Dogma is not the unique peculiarity of the Church. Every kind of systematic thought has to begin from some starting point. It has to begin by taking some things for granted' (8). 'The work of philosophers and historians of science ... has shown very clearly that the whole work of modern science rests on faith-commitments which cannot themselves be demonstrated by the methods of science' (20).

of meaning. Each of these two types of questions is then thought to involve its own unique approach. Undoubtedly, this is partially true. The natural sciences look for causal links between observational data, whereas theology is much more interested in linkages of meaning, which illuminate the significance of various things and events – a difference that naturally results in quite different research methods. Yet, on closer inspection, we discover that the difference at stake here is, at most, a difference of degree rather than a structural difference. If we consider the above list of twelve theological 'puzzles' once more, we notice quickly that they do not only concern questions of meaning. On the contrary, apart from questions 2, 3 and 10, all the questions are formulated as 'questions of fact'.

Yet we have seen what the status of facts is in the newer theories of science. Properly understood, the meaning of the term 'fact' has now been completely restored to its etymological origin. The word 'fact' derives from the Latin *factum*, i.e. the past participle of the verb *facere* = to make. A fact is, then, literally a creation or a product. Formulated more precisely: it is, as we have seen, a state of affairs *interpreted* in a particular way.[43] Without such an interpretation, we would not even know where a fact begins and ends, because we would be dealing only with a chaotic mish-mash of sensory impressions. Thus, as has been said since Kant, facts are by definition interpreted. That is to say: a particular meaning has already been attached to them. As a result, a strict separation between questions of fact and questions of meaning is impossible. Science too deals with questions of meaning; there too, explanation usually means not only discovering causal links, but also placing things within a larger framework, which provides us with insight. Scientists are therefore also often engaged in hermeneutics: Not only do they show that b is the result of c (for what would that mean divorced from any context?), but they also place b and c within a wider context that shows how the causal relation between the two can be *understood.*

In theology, this wider context expands, as it were, to the most comprehensive level at which the question of meaning can be posed: What is the meaning of reality as a whole, including our human existence? Yet, at the same time we should recognize that theology never concerns itself with providing meaning in a way that is independent of factual states of affairs. If, for instance, we were of the opinion that Jesus never lived at all (a view that, at one stage, counted as scientific, but is no longer taken seriously), then it would surely become quite

43 Brümmer defines facts as 'states of affairs which have been described and declared beyond doubt', *Theology and Philosophical Inquiry*, 240; it does not seem necessary, however, that a fact be *described* – unless, by description, one means interpretation. That it concerns a state of affairs *declared beyond doubt*, seems tautological, for when one speaks of x as a 'state of affairs', that already implies that one declares x beyond doubt.

problematic to define the meaning of reality and of our life in relation to Jesus, as Christians do. Thus the theologian is always *also* strongly interested in facts and their explanation, and cannot allow himself to retreat to a supposed terrain of pure meaning, where questions of fact are no longer at stake. What can be said, is that, in investigating and confirming such facts, the theologian makes use of methods and/or publications of other specialists, like historians, exegetes, and (depending on the subject that is being studied) also archaeologists, natural scientists and social scientists. Thus the theologian who concerns himself with Israel will have to relate in some way to historical data concerning the Jewish people, and the theologian dealing with creation to cosmological views concerning the origin of the universe. One need not become a historian or cosmologist oneself in order to do this in a professional manner.

Does the difference between theology and the other sciences then perhaps lie in the fact that theology appeals to revelation? This is also often raised as an objection to the scientific status of theology. Kuitert, for example, states unambiguously that, 'from the perspective of the philosophy of science, there is no place for revelation', and for this reason denies that classical theology is scientific theology.[44] His argument is that claims to revelation cannot be checked, and therefore go against one of the most important criteria for the practice of science. Yet it is questionable whether this is a valid argument. To be sure, claims to revelation do not allow for *straightforward* checks, by just testing them against reality in a kind of one-to-one relation (as 'is-sentences'). That is not how theology works – but, as we have seen time and again in the preceding chapters, nor is it how science works. In both cases, it is precisely the larger wholes (paradigms, or frameworks of interpretation, or whatever one wants to call them) that have to prove their worth; and they have to do so in life itself, so to speak.

Suppose someone believes on biblical grounds (so not because of arguments but based on revelation) that God exists. Such a belief is not, then, amenable to testing as an isolated claim, but rather within the meaning-giving context of the whole way of life by which it is implied, as Vincent Brümmer, among others, has argued in various places.[45] It is the form of life – let us say of living with God in faith – that does, or does not, prove its mettle in the experiences that I undergo in the course of my life. From that, it will transpire whether my faith (i.e. my claim to revelation) does more justice than various alternatives – either based on contradictory

44 H.M. Kuitert, *Filosofie van de theologie*, Leiden 1988, 17f., 31f. (the quotation on 18).

45 Cf. e.g. *Brümmer on Meaning and the Christian Faith. Collected Writings of Vincent Brümmer* (Ashgate Contemporary Thinkers on Religion), Aldershot 2006, 196, 207f., 314; truth claims about the existence and nature of God can only be tested within the context of 'the total fabric of religious beliefs in terms of which believers understand their lives and experience meaningfully' (208)..

claims to revelation or accounted for in some other way – to the whole set of relevant experiences that I undergo. Thus the test is, so to speak, *experiential* in nature.

The problem appears to be that Kuitert – and many others with him – can evidently picture an appeal to revelation only as an intellectualist quest for absolute, apriori certainty. This is indeed to some extent how the matter used to be approached in a large part of Kuitert's neo-Calvinist tradition. Its founding father, Abraham Kuyper, for instance, sought an epistemological guarantee, prior to everything, in the form of a speculative grounding of the authority of Scripture that excludes every possible type of doubt.[46] This neo-Calvinist version of foundationalism in theology has (as can be witnessed in Kuyper's spiritual descendants) proved just as untenable in theology as its neo-positivist counterpart in the philosophy of science. Yet there is an alternative. An alternative approach *must* perhaps even be adopted, for, as we have seen, even in the sciences we cannot avoid some kind of appeal to revelation. That is to say: in scientific research too, we deal with insights that precede everything and seem to us undeniable, but the testing of which is not all that simple, and in the first instance (or perhaps even permanently) stands out. We saw that Lakatos situated insights of this kind in the 'hard core' of his research programmes, and that Kuhn's paradigms also stem from such fundamental guiding insights. In this sense, an appeal to revelation can, as indicated,[47] definitely not be ruled out-of-court from the perspective of the philosophy of science, not only in the context of discovery (the *Aha-Erlebnis* of the scientist enjoying a *eureka*-moment), but also in the context of justification – for as we have seen, these two cannot be so neatly separated.

Thus theology does nothing unusual by appealing to revelation in this way, as also appears from the fact that the importance of a similar appeal to what is granted to us from beyond as a gift is again being recognized in contemporary continental phenomelogy.[48] Only, we cannot theoretically safeguard such an appeal against all doubt and criticism at the outset, for instance by means of a certain theory of inspiration (such a theory has its legitimate place among the

46 Cf. on this Dirk van Keulen, 'The Internal Tension in Kuyper's Doctrine of Organic Inspiration of Scripture', in: Cornelis van der Kooi and Jan de Bruijn (eds.), *Kuyper Reconsidered. Aspects of his Life and Work*, Amsterdam 1999, 123-130. 'Kuyper is searching for a form of certainty that is absolute and excludes all doubt – a certainty that can never be challenged. (...) the certainty of faith he was looking for was *rational* in character' (129).

47 Above, section 4.3.

48 Cf. e.g. J.D. Caputo & M.J. Scanlon (eds.), *God, the Gift, and Postmodernism*, Bloomington 1999; D. Janicaud et al., *Phenomenology and the Theological Turn. The French Debate*, New York 2000; Jean-Luc Marion, *Being Given. Toward a Phenomenology of Givenness*, Stanford 2002.

auxiliary hypotheses rather than in the hard core of a theological programme). A scientific theology would, then, also have to account for its appeal to revelation. This is, moreover, in complete accordance with its goal, for its duty 'to make [a] defence to anyone who demands from you an account of the hope that is in you' (NRSV)[49] does not cease as soon as we come to the belief that God has *revealed* himself. However, such an account should not have a foundationalist character, as if the theologian were called to establish, by his own powers, a solid base of rational arguments underneath the belief in divine revelation (or, for that matter, underneath the Bible). Rather, such an account will usually consist in the acknowledgement that one has simply been convinced by the *content* of the biblical message, and in a narrative about how and why that took place. For that content indeed belongs to the heart of the paradigm. Next to this, theology will attempt to show the *coherence* that faith brings to one's life. The faithful too have a 'web of belief' (Quine) in which some elements are more central than others, and in which they continually seek to maintain a relation with the comprehensive totality of their convictions (of whatever kind). Here, Lakatos' ideas concerning a 'hard core' and a 'protective belt' are relevant. Rather than a vertical structure, where everything rests on the weight of a single fundamental conviction, our network of beliefs has a horizontal structure, where things mutually cohere in all sorts of subtle ways, and mutually support one another.[50]

Is this a way of rendering one's faith unassailable, of erecting a shield against all possible criticism, of immunising oneself against any form of proper testing? No, for this faith, like all other paradigmatic networks, will have to prove its integrating power – its 'puzzle-solving capacity' – in everyday life. Here too, the rule applies: The proof of the pudding is in the eating. Just as in 'normal science' (Kuhn) a scientist working within a particular paradigm or research programme is deeply convinced of its correctness (how often did we see Newtonian mechanics not being verified?), a theologian will also usually be quite confident that his deeply rooted faith will be vindicated by this testing. Ultimately, God himself will see to that ... Nevertheless, there may be periods when paradigms face a hard time, and research programmes seem to degenerate. During such times, there will be doubts, and the realisation will dawn on us, quite rightly, that all human knowing is fallible. In this manner, faith also sometimes resembles a degenerating programme: It is also occasionally *not* confirmed for such a long time that questions

49 1 Peter 3:15; in the Petrine epistles, the term 'hope' is often *pars pro toto* for the whole Christian faith. See e.g. I. Howard Marshall, *1 Peter*, Leicester 1991, 35, 115f.

50 See for a development of this insight Paul Helm, *Faith with Reason*, Oxford 2000, especially 43-65.

start to arise about its sense of reality. In fact, faith is already pictured in this way in the Bible (think of the Book of Job, of various Psalms etc.).

Thus faith is structurally vulnerable. It lies exposed to forms of doubt and uncertainty that can be called forth or exacerbated by various events and developments.[51] In present-day Western Europe, we can think, for instance, of secularisation and growing religious pluralism. Today, there is therefore greater need of a theology of the cross, which is familiar with perseverance in the midst of doubt, than of a theology of glory, which believes itself victorious over all doubt. After all, the latter type of theology may soon turn out to be untenable. We need, after all, to be conscious of the human – and therefore fallible – character of all our knowing. It is with good reason that this 'fallibility' is amongst the most uncontested results of the newer philosophy of science.[52] If we believe that, in other respects, the newer philosophy of science opens up new perspectives for theology that have so far been too long neglected in theological reflection, then we should also take this flipside seriously. As far as theological content is concerned, this means, for instance, that we should not (consciously or unconsciously) understand the doctrine of the *perseverantia sanctorum*, the so-called 'perseverance of the saints', as a doctrine about the inherent self-confidence and unquestionability of belief, but as what it originally was: a statement of *faith*. We are then fully conscious of our own fallibility, yet, on the basis of experiences which we have undergone, express trust that, in the future, *God* will confirm the belief in Him.

I conclude, then, that, in this regard too, there is no *difference in principle* between a theology informed by faith on the one hand, and the (other) sciences on the other: in both cases, we cannot but speak of revelation (in the sense of insights from beyond, which cannot be simply proved, but which nevertheless impress themselves on us with such force that we feel inwardly certain of them[53]); in both cases, we depend on *givens* in the most literal sense of the word: things that are given to us; but in neither case does this mean that we are exempt from the duty to give an adequate account, or from the need for future confirmation, for our knowledge is always 'in part', that is to say, fragmentary and vulnerable.

51 Cf. A. van de Beek, 'Is God Detectable?', in: Henri A. Krop et al. (eds.), *Post-Theism. Rephraming the Judeo-Christian Tradition*, Leuven 2000, 209-221 on 'the contested position of the believer' (220).

52 See above, section 2.2, on Karl Popper's 'fallibilism'; by the way, the term fallibilism was already introduced into philosophical discourse by American philosopher Charles S. Peirce.

53 Cf. in this connection the way in which John Calvin spoke of the 'internal testimony of the Holy Spirit' in relation to the authority of Scripture. See Henk van den Belt, *The Authority of Scripture in Reformed Theology. Truth and Trust*, Leiden 2008, chapter 2.

Nor can it be maintained, then, that theology necessarily places itself beyond the rules of the game of the scientific community when it appeals to revelation in this way – not even when it appeals to 'special revelation', for revelation, correctly understood, is always 'special'.[54] Here too, no structural difference can be identified between theology and the (other) sciences (e.g. the natural sciences). So, when Arie Molendijk concludes that the *parti pris* or bias of theology seems irreconcilable with public scientific research, this is a questionable conclusion, to say the least. The idea that a certain bond with church and dogma belongs to the discourse of systematic theology is true; to suggest that this involves one of the greatest differences from modern science, however, is to ignore the paradigmatic nature of all scientific research. The way in which a theologian feels bound to a particular confession is not, in principle, different from the way in which a natural scientist (or social scientist) works within a particular given paradigm.

In the final analysis, then, I see only one relevant difference between the two domains, albeit one which is more a matter of degree than structure: in theology, the number of paradigms competing with one another seems *much larger* than in the natural sciences. Here, we should, of course, attend carefully to the definition of the concept of a paradigm. Since Kuhn, this concept has been subject to enormous devaluation, as a result of which, even in the case of a theory that differs only slightly from an earlier theory, a 'new paradigm' is quickly spoken of. Kuhn, by contrast, understood paradigms as *incommensurable*, that is to say: the difference between them is so great that, from the perspective of one's own paradigm, it is almost impossible to understand someone living in another paradigm. We must first learn to speak that person's *language* – the parallel with the demanding and drawn-out process of learning a second language has in fact been drawn by Kuhn.

Seen from this perspective, the number of paradigms in the world of religion and theology is not all that large. It is debatable, for example, whether we ought to speak of a Protestant as well as a Roman Catholic paradigm, or even, like Küng, of the paradigm of Luther as well as that of, say, Calvin. Luther and Calvin could, after all, understand each other very well, and generally speaking, Protestants and Catholics are also able to make clear to one another the points on which they hold different views. Naturally, moments may arrive when it proves impossible to convey one's deepest feelings to the other, but moments like that presumably also occur whenever people who share the same religious faith engage in deep conversation with one another. It is advisable, therefore, not to attribute a distinctive paradigm to every current or community within Christianity, but to speak of *the*

54 Hendrikus Berkhof, *Christian Faith. An Introduction to the Study of the Faith*, Grand Rapids 2002, 82.

Christian faith as a paradigm – a paradigm that is then systematically studied and reconstructed by theology. For clearly, on this level it can be justifiably argued that for others to come to understand what this paradigm is all about will require a long process of familiarization, just as, vice versa, the same applies to Christians with regard to, say, the Buddhist paradigm.

Yet, even if we define the concept of a paradigm in this way, the number of mutually exclusive paradigms in the world of faith and theology is still much larger than in the natural sciences. We at least have to attribute a distinctive paradigm to each of the great world religions, whilst in addition it seems that the different varieties of atheism (such as humanism, which attaches great value to human freedom, and scientism, which in fact denies that very freedom) also struggle to comprehend one another. And then we have not even yet spoken of the in-between varieties, like the so-called 'ietsism' ('Do you believe in God?' 'No, but there must be something!')[55] or the post-modern fashion to combine, with a *pick-and-choose* mentality, loose fragments from a variety of spiritual traditions – varieties that usually have yet to prove their viability. In the natural sciences, however, one paradigm tends to dominate; it is only in situations of crisis that competition becomes an issue, and even then, the number of alternatives usually remains very limited. To be sure, many competing research programmes run simultaneously in various fields, but they do not differ from one another in a paradigmatic sense: there is basic agreement on many relevant questions. The amount of what Willem Drees calls 'consolidated knowledge'[56] is thus, on the whole, much higher than in the area of views of life and theology.

There is only one plausible explanation for this difference: the much higher degree of *complexity* of the questions at stake in theology. After all, faith and theology are ultimately concerned with the meaning of reality as a whole. Hence theology cannot limit itself, in its investigations, to a particular segment of reality, but is faced with the enormous task of making sense of *all* our different types of experience (a task it shares, incidentally, with philosophy). The theologian does not have to practice various methods of other sciences, it is true, but must be able to have some grasp of their implications and results. At the very least, theology involves creatively relating historical, exegetical and hermeneutical methods to one another. In many cases, other sciences also play a role in addition to these

55 Ietsism (literally: 'something-ism') is a recent loanword from the Dutch. According to *Wikipedia*, it stands for a range of beliefs held by people who, on the one hand, inwardly suspect – or indeed believe – that there is „more between heaven and earth" than we know about, but on the other hand do not necessarily accept or subscribe to the established belief system, dogma or view of the nature of God offered by any particular religion' (http://en.wikipedia.org/wiki/ietsism; last visited 30 June 2009).

56 Willem B. Drees, *Creation*, 66.

(such as the natural sciences in the doctrine of creation, the human sciences in theological anthropology, et cetera).

Recalling Dooyeweerd and Vollenhoven's theory of modal aspects, we see that the 'higher up the ladder' we climb, the greater the number of competing schools of thought in the disciplines connected with the various aspects. For example, there is evidently more paradigmatic plurality in economics than in biology. Perhaps we might say that theology reaches the limit in this regard. This means that theological hypotheses are often very complex, and far less amenable to precise formulation than hypotheses in for example the natural sciences. Consequently, the testing procedures in the theological field are, in turn, much harder to implement, and, at the very least, involve a much longer haul. Thus we could say that it took many centuries before the worship of Zeus became falsified, as it became clear that it was based on a faulty paradigm. Yet various other mutually exclusive theological paradigms continue to exist side by side.

Apparently, it belongs to the human condition – and, here, a Christian would undoubtedly see one of the tragic consequences of sin – that we cannot convince each other (even as people who are reflective, intelligent, well-educated, et cetera) of the truth or falsity of answers to the deepest questions of life. Of course, it does not follow that, for example as a Christian, one cannot be convinced *oneself* of what one believes.[57] Yet the implications are far-reaching, for on one thing we would all agree: The questions at stake here are not exactly of little importance. On the contrary, it makes quite a big difference whether our reality is 'intended' or merely a coincidental fall of the dice (cf. above, question 2). And it really matters whether or not the forces of chaos, suffering and death have the final say in world history (question 12; the same can be argued with regard to all the other questions). In light of this, one would expect the disciplined study of these questions in theological and philosophical educations to be very popular!

However, one of the most remarkable features of our culture of 'apatheism'[58] is in fact the incredible indifference it exudes with regard to questions like these. Of course this indifference as to what should be of ultimate concern to us can be explained to a large extent by cultural factors (e.g. post-modern scepticism with regard to all grand narratives, the widespread hedonism in Western culture, the rise of a 'sensation culture' revolving around short-lived hypes and transitory emotions, et cetera); but that is not to say that it is justified. René van Woudenberg has

57 Cf. Wolterstorff, *Reason within the Bounds of Religion*, 58: From the collapse of foundationalism '... it does not follow that the Christian cannot have any confidence in what he believes. Nor does it follow that none of what the Christian believes has the status of knowledge'.

58 See above, chapter 1, footnote 2.

compared the message sent into the world by Christianity to a bomb threat. It is a rather menacing metaphor, yet, at least in one respect, it is illuminating. Even if we have no idea whether the rumour is true, its implications are so overwhelming, and the consequences involved so far-reaching, that it should be taken seriously anyhow. The fact that the security officer is unsure whether the telephoned bomb threat he received is based on truth, is no reason for him not to take it seriously. The stakes are much too high for that.[59]

Likewise, the great variety of philosophical and theological orientations should not seduce us into consigning this whole area to the private sphere. After all, even from a purely cultural perspective, it is in our common interest to have centres – academic or otherwise – where 'trans-communal' criteria are sought with reference to which a discussion about sense and nonsense in religions and views of life can be conducted. But in addition to this, from a scientific perspective the great variety of views of life should be seen as a challenge to embark upon intensive study and research. After all, science is, by definition, oriented towards the unknown, towards that which is not (yet) certain. It is not rational, precisely when it comes to the ultimate questions in life, to assume in advance that nothing sensible can be said about truth and falsity in this area, that we cannot know anything about God et cetera. It is clear, in any case, that for a university to do this (and to show this by, for instance, closing its theological faculty, or breaking it up into modules), is to opt for limiting itself to less important questions, thereby sacrificing part of its public relevance. For clearly, since matters of truth are at stake, religion and theology are a matter of public concern.[60]

The Christian community also has a clear interest in public reflection on the central aims and tenets of religions and other views of life. After all, Christian faith is not intended to serve the special interests of the limited community of believing individuals, but has a wider range. Only a sect would regard that as satisfactory. Now that the difference between a church and a sect can increasingly no longer be defined quantitatively (as if a church were big, and a sect small), we may perhaps find here a difference that is based much more on principle: a church is not only concerned with its individuals, but also with the society and culture that surround it. It is concerned with them in a dual sense, namely in that it truly wants to listen and learn, but also in that it has something to say to them. It wants, in a particular way, to remain on speaking terms with the culture, and with the intelligentsia who play a leading role in it. In this sense, its theology is

59 René van Woudenberg, *Toeval en ontwerp in de wereld. Apologetische analyses*, Budel 2003, 5f.

60 For this and a number of other arguments for this position, see Roger Trigg, *Religion in the Public Life. Must Faith be Privatized?*, Oxford 2007.

missionary, and not only in its own interest. After all, it is in everyone's interest that more clarity be achieved as to where we collectively stand, as to whether or not reality is dependent on a God (and if so, which God), whether the Gospel indeed possesses the 'problem-solving' and meaning-giving capacity that many people all over the world ascribe to it, et cetera.[61]

In this way, our reflection on the relationship between theology and science results in a plea for the restoration of the public function of theology. It will be clear that it is not possible to do so in a way that transcends paradigms. Therefore, the idea once put forward by Hendrikus Berkhof of a *multiplex ordo* in a theological-cum-philosophical faculty (housing e.g. Christian, Jewish, Islamic and humanist studies), does not seem all that far-fetched from the point of view of structure.[62] The disciplines and methods traditionally taught in the faculties would then be used as building blocks with which, ultimately, the question of truth would be opened up for discussion. Only then could we speak of a truly *theological* faculty.

Needless to say, the idea behind this is not to have just any paradigm whatsoever represented here, but at least those paradigms that are among the most important options in society. In the Western countries, these include, apart from the Christian paradigm, the scientistic and classical humanist paradigms. Although there are important similarities between these two, it seems advisable to distinguish them from one another, given the totally different values that they attach to human beings. The percentage of Muslims in the makeup of the population in Western Europe and the United States is often vastly overestimated. The very idea that Western society is 'multi-religious' is quite exaggerated, and largely based on a secular myth. There is really only one religion by which Western society has been deeply shaped and influenced, even to this day, and that is the Christian religion. Nevertheless, it will certainly contribute to the common good that, along with the rise of Islamic studies, also the Muslim paradigm is welcomed today at a number of universities. However one may wish to determine the number of vital paradigms, much would already be achieved if academic theologians

61 Some ideas unfolded here are inspired by Dutch theologian A.A. van Ruler (1908-1970), who considered theology as a function of the state rather than the church, and emphasized that the Christian message is intended to renew not only individual persons but also cultures and societies. In what is to become a huge project Van Ruler's collected writings are now being (re)published in a critical edition. See Dirk van Keulen (ed.), *Verzameld werk* 1 (De aard van de theologie), Zoetermeer 2007.

62 Berkhof, 'God: voorwerp van wetenschap? II', in: idem, *Bruggen en bruggenhoofden*, 215: 'If we want to keep our rapidly pluralising country religiously habitable, we should push ahead towards a multiplex ordo for the 21st century'. For a slightly different proposal as to how the the public function of theology might be given concrete form (viz. in terms of a 'Christian university'), see Gavin D'Costa, *Theology in the Public Square*, Oxford 2005.

(including those in the state-funded faculties) were to make their paradigmatic views of life explicit, and relate them to their work. That would, in any case, help to remove the misunderstanding that it is possible to do theology in a neutral fashion.

If a *multiplex ordo* ever became a reality, the paradigms represented in it would, on the one hand, get the opportunity to elaborate themselves by continually coming to terms with current developments (e.g. in the sciences). On the other hand, 'trans-communal' criteria would have to be sought from within the diverging paradigms, with reference to which a degree of mutual understanding could at least be achieved, and a discussion embarked upon concerning questions such as 1 to 12 above. In such a setting, the humanist could engage in discussion at the cutting edge with the Muslim theologian, and the Christian theologian would have to face critical questions, in a spirit of intellectual integrity, from the adherents of scientism (and vice versa), et cetera. The fact that they would, presumably, not reach agreement easily, should not discourage us from creating this sort of platform. It is, for better or for worse, in society's interest that, amidst the process of the fragmenting and polarisation of different (often ideologically-determined) groups, an open space be reserved where channels of conversation can be kept open right down to the deepest level.

The suggestion that, as far as their view of life is concerned, a large majority of people have more or less the same beliefs, and that only some isolated and increasingly inaudible deviant voices are still heard from the more backward sectors of society, has, after all, proved an outdated myth of modernity. At best, it is 'also a kind of belief'. The rise of Islam in Western Europe alone will eventually make it increasingly difficult to ban religion from public discourse in favour of the lowest common denominator. Opening up the question of truth *in religiosis* for discussion is therefore in the public interest. Whenever a 'top-class university' comes into existence, it would surely also deserve a theological faculty. For, apart from the scientific legitimacy that we have found, there is also an urgent social need to pay close attention to theology in the original sense of the word. This can best be achieved by granting theology its rightful place in the public university. In this way, theology itself can, once again, give effect to its threefold orientation – namely, towards the church, academia and society[63] – and avoid being marginalised to the position of an affair internal to the churches. In this way, it will remain what it has been from the very beginning: a public affair, between faith and science.

In the final section of this book, I would like to bring yet more focus to our thoughts on this threefold orientation of theology. But first, it is now time to shift

63 On this triad, see David Tracy, *The Analogical Imagination. Christian Theology and the Culture of Pluralism*, New York 1981, and cf. below, section 8.2.

our gaze from theology to the contents of faith. When we subject the contemporary debate on (and between) faith and science to closer investigation, do we then find our conclusions with regard to the importance of theology confirmed? Or does a picture then emerge that is structurally different after all, for instance because the (natural) sciences have long since lost all possible interest in notions like faith and theology? Let us examine this in the next chapter.

7. Religion and science

7.1 Ian Barbour's classification

The field of questions regarding the relationship between religions and views of life on the one hand and the sciences on the other is sometimes referred to as 'faith and science', or, more commonly, 'science and religion'. The suggestion that both pairs of concepts give rise to is that faith or religion and science in one way or another form categories that are comparable with each other, and that it is obvious that we should relate them to each other by means of these tags.

However, the question is whether this really is so obvious. There is much to be said, as we have seen in the previous chapters, for comparing *theology* and science with one another. For clearly, theology and science are both in a certain way second-order activities. The sciences abstract from the direct sense experiences that we undergo and try to isolate, order, analyse, explain etc., these experiences.[1] Theology does something similar with respect to the reality of faith. It abstracts from the religious experiences that people have had (or say that they have had), and tries to order these experiences and to evaluate them.[2] So, both theology and the (other) sciences take, as it were, a step back, in order to observe their object of study from a certain distance. Categories such as faith and religion by contrast are much more closely connected to everyday experiences and practices. They are related to a whole complex of often spontaneous activities, which usually are not as rational and methodical as what we do in the sciences and theology. For this reason it is more natural to compare theology and science with each other than faith and science.[3]

1 In relation to our findings in chapter 3, we are now generalizing here somewhat; strictly speaking sensory experience does not play a role in all sciences, as we saw for example is the case with mathematics.

2 For the parallel between sensory and religious experience, we again point to William Alston, *Perceiving God. The Epistemology of Religious Experience*, Ithaca 1991. For a critical discussion of Alston's work, see Wessel Stoker, *Is Faith Rational? A Hermeneutical-Phenomenological Approach*, Leuven 2006, 47-82.

3 Accordingly, in the English-speaking world we see this happening more and more, as is clear from the increasing popularity of expressions such as 'theology and science' or 'science and theology'.

Yet the latter is what we shall do in this chapter, and with good reason. For what is going on in the sciences is not only of concern to professional theologians, but no less to ordinary believers. Moreover, we have seen that theology and faith are very closely linked to one another, at least given the definition of theology which we used as our starting point in the previous chapters. Theology – in any case systematic theology – is the disciplined reflection on the nature and content of faith. It aims to give an account of the more or less spontaneous belief statements and other utterances of faith and trust that come from individual believers and communities of believers (for example in the churches). Because theology and faith are so closely related to each other, it is also possible to relate religion and science to each other and to compare them with each other. For clearly, both are practices in which in some sense truth claims (or at least factual statements) are being made. Therefore, in such a comparison we especially have to consider how particular statements of faith – such as 'God created heaven and earth' – relate to the results (or supposed results) of scientific research.

There is yet another reason, however, why tags like 'faith and science' or 'science and religion' are in fact inappropriate, and that is because these terms are invariably used in the singular. We have of course seen that there is an enormous variety of sciences (chapter 3), whilst it does of course make quite a difference as to exactly what sort of faith or religion we are dealing with. If we are dealing with Christian faith – and again it is to this that we shall limit ourselves here – then traditionally the results of research in the *natural* sciences and more specifically in physics and biology, are of central concern. One only has to think of the names of figures who, from a historical perspective, were seminal in this regard such as Galileo Galilei and Charles Darwin. But we do not have to limit ourselves to the natural sciences when studying the interaction between Christian faith and the sciences. It can also be extremely useful and instructive to consider the results of, for example, psychological or sociological research for their implications for the Christian faith or for religion more generally. And if we look at the natural sciences, we do not have to restrict ourselves merely to classical disciplines such as physics and biology. Perhaps the recent discoveries by the so-called neuro-sciences concerning the workings of our brains are much more intriguing and also much more problematic for faith than the challenges of the traditional natural sciences.

It will be clear that in doing this a really wide range of questions and subjects arise which, moreover, will all demand a certain level of specialist knowledge. It will therefore be impossible to go into everything that requires our attention here. Rather than concentrating on – and as a result probably also losing ourselves in – all sorts of substantive questions which arise concerning the relation between religions and the sciences we want to look at what sort of *basic views* there are about this relationship and what the pros and cons of each of these are. In so doing, we shall perhaps be able to gain an overview of the subject as a whole, while using some important substantive questions for the purposes of illustration. For reasons of brevity, we shall continue to indicate the field by means of the phrase that is still most common today: science and religion.

When it comes to basic views on the relationship between science and religion, we are greatly helped by an extraordinarily influential classification which was developed by probably the most frequently-cited modern writer in the field, the American natural scientist and theologian Ian Barbour. In a number of publications, Barbour has distinguished between four ways in which science and religion have been related to each other in the past and continue to be related to each other today.[4] He denotes these four ways respectively as *Conflict*, *Independence*, *Dialogue*, and *Integration*. We can see science and religion as being continually and by definition in conflict with each other; or we can see them as two separate worlds, which, due to this difference in category, can in no way clash with each other; or we can identify certain common ground or analogical similarities between them which allow for dialogue; or we can try to unite both of them in one overarching concept. With these four options the field of possibilities is not fully exhausted, but it still seems that these four are the most important ones.[5]

The advantage of this classification (or taxonomy, as Barbour himself calls it) is that many positions in respect to the relationship between science and religion which at first sight seem very different, can be classified by means of what turn out to be fundamental structural similarities. An extremely complicated, polymorphous whole can thus be reduced to a limited and manageable number of basic positions, that we could refer to as *models*. This does not imply that things are oversimplified because Barbour does not refrain from pointing to the important underlying mutual differences between various theories and positions that he subsumes under one and the same model. Although other classifications do exist,

4 Ian G. Barbour, *Religion in an Age of Science*, San Francisco 1990, 3-30 (revised as *Religion and Science: Historical and Contemporary Issues*, San Francisco 1997); repeated in *When Science Meets Religion*, London 2000 and summarized in his entry 'Science and Religion: Models and Relations' in J. Wentzel Vrede van Huyssteen (ed.), *Encyclopaedia of Science and Religion*, New York 2003, 760-766 (these are only some of the at least seven scholarly publications in which Barbour has put forward his typology).

5 In my paper 'Geloof en wetenschap: op zoek naar consonantie', in: Cees Dekker, René van Woudenberg & Gijsbert van den Brink (eds.), *Omhoog kijken in platland. Over geloven in de wetenschap*, Kampen 2007, 50-75, I distinguish two other possible models: science and religion may just touch each other like two adjacent circles, or the one may comprise the other (which means that science totally determines the content of religious faith, or vice versa)…

which each have their own advantages, we shall focus in this chapter on Barbour's classification, which is still one of the simplest and most convenient. We shall not do this, however, without first making some preliminary comments in relation to the classification. For there is some reason to criticize the way in which Barbour sets about his task.[6] In my opinion, this does not mean that we should reject Barbour's classification, but rather that we ought to be aware of some of the guiding presuppositions which lie behind it.

First of all, there is clearly a rhetorical element in the names that Barbour gives to his four models. Well-educated Western people have a dislike for conflicts, and so, given the four possibilities that Barbour provides, we shall not immediately tend to choose the conflict model between science and religion. 'Independence' too can sometimes have negative overtones, carrying with it the idea of breaking away and separation. Usually we prefer to manage things differently – for example by having a dialogue with each other. Dialogue is almost a buzz-word and anyone who tries to dialogue with anyone or anything else can almost automatically count on general approval. Finally, integration is also seen as an important, positive value, as we know all too well since an 'integration debate' has sprung up everywhere regarding the settling of immigrant populations in West European and American societies.

So there is a clear direction in Barbour's classification.[7] But fortunately Barbour does not make any bones about this. He makes it quite clear that it is his intention to convince the reader that the conflict and independence models are inadequate and that we must strive after dialogue and (in certain circumstances) integration between science and religion.[8] And so we are left to judge Barbour's arguments on their merits. In doing so, we should be aware that it is also possible that the dialogue model will turn out to be less fit-for-purpose than for example the conflict model. Moreover, we must not necessarily go along with the order in which Barbour presents his four models. For underlying this there is a sort of idea of progress, whereby we gradually emerge from the first stages of conflict and separation in order finally to reach integration having gone through dialogue. But if indeed such a development might be proven from a historical perspective

6 Here, I only call attention to the most important ones. In my contribution mentioned in the previous footnote, I briefly discuss five objections that are put forward against Barbour's typology in the literature (55-60).

7 On this point, see also Geoffrey Cantor & Chris Kenny, 'Barbour's Four-fold Way: Problems with His Taxonomy of Science-Religion Relationships', *Zygon* 36 (2001), 765-781 (esp. 766); and more generally Christian Berg, 'Barbour's Way(s) of Relating Science and Theology', in: Robert J. Russell (ed.), *Fifty Years in Science and Religion. Ian G. Barbour and His Legacy*, Aldershot 2004, 61-75.

8 Barbour, *Religion and Science*, 77.

(which is not at all the case), then it could still be so that the later phases signify a regression, for example with regard to the independence model. This being so, it will be good to approach the various possibilities in as open a way as possible.

Secondly, the same applies to Barbour's models as it does to all models: they draw our attention to certain similarities and differences, but in doing so they mask other similarities and differences.[9] The suggestion then is that these other similarities and differences are of secondary importance. But this is of course based on taking a certain perspective and in a certain sense it reveals which values we find important. In this connection, Willem Drees is right in saying that on the one hand Barbour's classification lumps everything together that is in fact very different, whilst at the same time differentiating between positions which are in important respects closely related. As an example of the second case, he gives so-called scientific materialism, which Barbour classifies under the conflict model, but which at the same time very much looks like an attempt at a metaphysical integration of science and religion, in which both are woven into a whole. In my opinion Barbour could adduce against this that hardly any significance is ascribed to faith in this overarching materialist vision, but this does not take away the fact that Drees draws attention to an important risk that attaches to Barbour's classification.[10]

However, as we have said, we do not think that Barbour's classification ought to be discounted for this reason. After all, in every classification certain similarities and differences are emphasized and others are not. We cannot, as we have already seen in chapter 4, get away from a certain perspective, or from making certain choices. Alternative classifications of the possible relationships between science and religion necessarily bring with them other limitations, and could put us on the wrong track again in different ways. But this is not to say that they are better or worse, rather that classifications such as these always have a limited significance. Instead of using this as a reason for leaving them aside, however, we can try to make ourselves aware of the limitations of a classification such as Barbour's, in order to profit nevertheless from the clarification that it brings to an extremely opaque whole. With these qualifications in mind, we now want to look more closely at Barbour's models and provide a commentary on them.

9 For an illuminating account of the nature of models, their relationship to metaphors and their function in science, religion and theology, see Vincent Brümmer, *The Model of Love*, Cambridge 1993, 3-29.

10 Willem B. Drees, *Religion, Science and Naturalism*, Cambridge 1996, 43.

7.2 The conflict model: science and religion at war

The idea that faith and science are locked into a long-lasting conflict with each other is still deeply embedded in the whole of Western society. Of the four models that Barbour puts forward, this is without doubt the most popular amongst the population at large. Whether our sympathy is more with faith or with science, collectively we have the impression that faith and science are continually encroaching upon one another's territory, much to their mutual annoyance. Believers often see the march of science as a threat to their religion, and supporters of scientism see faith as an irritating and irrational obstacle to the dissemination of their enlightened ideas.

From a historical perspective, there is of course also good reason to see the relationship between faith and science as one which is full of unease and conflict. Interestingly enough, however, the use of the metaphor of conflict in this connection is not so very old. In fact, it only emerges at the end of the nineteenth century, when two influential studies described the history of the relationship between science and religion as the history of a constant struggle. A good indication of this is the title of the two books. John William Draper called his book *History of the Conflict between Religion and Science* (1875) and Andrew Dickson White's book had an even more dramatic title: *A History of the Warfare of Science with Theology in Christendom* (1896).[11] On closer inspection, however, it turns out that neither Draper nor White thought in such black and white terms. Draper aimed his attack almost exclusively at the Roman Catholic Church, because it had so often stood in the way of the free practice of science and the dissemination of its results, using its position of power. And for White the problem above all was with theologians, many of whom in his opinion wrongly concerned themselves with science instead of dealing with the results of historical biblical criticism. Neither Draper nor White was averse to religion and faith as such.[12]

Nevertheless, both authors did their best to portray the history of the interaction between religion and science as a history of senseless opposition from the church and theology to a science which would always turn out to be the winner in the end. In this way both of them have given a huge amount of fuel to the conflict model and have exercised an influence that it is difficult to overestimate. Meanwhile, modern-day historians of science are almost unanimous in holding

11 Both books were published in New York; the two-part work of A. D. White was in a certain sense a follow-up to and further development of his *The Warfare of Science*, New York 1876.

12 Claude Welch, 'Dispelling Some Myths about the Split between Theology and the Sciences in the Nineteenth Century', in: Richardson & Wildman (eds.), *Religion & Science*, New York 1996, 29-40. For the line which White was fighting against, see Drees, *Religion, Science and Naturalism*, 67f.

that in reality the situation has been much more complex. The reception, for example, of Darwin's ideas in the latter part of the nineteenth century was very varied: initial fierce theological opposition was succeeded by an enthusiastic reception of his ideas by theologians such as Daniel Friedrich Strauss, whilst other theologians tried to mediate between Darwin and orthodox Christianity. So the idea of a permanent conflict in no way tells the whole story.[13]

Next, if we have a look at another famous text-book example of a historical conflict between science and religion, the Galileo-affair, it turns out here too that the conflict thesis is hardly applicable. In the course of historical research it has become increasingly clear that the battle involving Galileo was not a conflict between or about science and religion. Rather it was an internal theological dispute over the question of who had the right to establish the authoritative explanation of the Bible. The Roman curia, which in response to the Reformation had dug its heels in on this point at the Council of Trent (1546), continued to insist that *the church* had this authority. In contrast, Galileo could be seen as a sort of crypto-Protestant at this point, believing that people could interpret the Bible for themselves, or at least could (and might) explain it in such a way that it did not come into conflict with the scientific discoveries of Copernicus. It is interesting to note that in this situation Galileo had support from amongst the representatives of the church, but also that he had opponents from amongst his fellow scientists! In the trial of Galileo as well there was only a question of authority: Did Galileo comply with the Papal ordinance of 1616 not to teach or even discuss the prohibited Copernican ideas? How could Galileo obtain an *imprimatur* (permission to print) for his *Dialogue on Two Chief World-Systems*, which appeared in 1632 and in which he treats the Copernican issue as more than only a hypothesis?[14] In short, the conflict model is far too static and monolithic to cover the complex historical reality of the various interactions between science and religion. Even in the most famous conflict-cases it turns out to be incomplete and misleading.

The fact that someone such as White was sometimes wide of the mark in his attempts to invoke this idea of a perpetual conflict can be seen from what he writes concerning the reaction of John Calvin to the Copernican system. He wrote:

Calvin took the lead, in his *Commentary on Genesis*, by condemning all who asserted that the earth is not at the centre of the universe. He clinched the matter by the

13 See e.g. Welch, 'Dispelling', 31; David N. Livingstone, *Darwin's Forgotten Defenders. The Encounter between Evangelical Theology and Evolutionary Thought*, Edinburgh 1987; James R. Moore, *The Post-Darwinian Controversies*, Cambridge 1979.

14 See the summary of the current situation in relation to research on Galileo by Ernan McMullin, 'Introduction', in: Ernan McMullin (ed.), *The Church and Galileo*, Notre Dame 2005, 3-7.

usual reference to the first verse of the ninety-third Psalm, and asked, „Who will venture to place the authority of Copernicus above that of the Holy Spirit?"[15]

It is interesting that White does not quote his source here. If he had tried to find the source, then he would have discovered that there is none. It seems that White copied this quotation without checking it from a book which appeared ten years earlier, written by a British theologian who was not altogether unknown at that time, F. W. Farrar. Farrar gives the same quotation, as E. Rosen has shown,[16] but where he took it from is still unclear to this day. In any case it was not from the work of Calvin, as Rejjer Hooykaas had already shown a few years previously.[17]

Hooykaas argued that the absence of this quotation from Calvin's works is not strange, since it would go against Calvin's exegetical principles to condemn Copernicus. Calvin's doctrine of accommodation (that is, his view that the Holy Spirit often adapts or accommodates himself in the Bible to the everyday language and thought processes of the readers in order to speak as comprehensibly as possible) made it unnecessary for him to read Ptolemy or, for that matter, Copernicus into the Bible – something which he therefore did not do. Anyone who wants to come to know more about astronomy must look elsewhere and not in the Bible, Calvin wrote in his commentary on Genesis 1:15. In this connection he recommended the study of astronomy (along with medicine and other natural sciences), as these lead us deeper into the wisdom of God in his creation.[18] Calvin himself was not an expert in this field, and so he just accepted that the Ptolemaic world view which had been held for many centuries was correct. As it happens, many intellectuals did the same in the sixteenth century, including astronomers, simply because the precise status of the Copernican system was not yet clear – the battle between the paradigms was still on-going.[19]

15 White, *History of Warfare,* part 1, 127. Incidentally, Psalm 93: 1 does not say that the earth is in the centre of the universe but only that it is established and shall not be moved. But some people have indeed deduced from this that the earth does not move round the sun.

16 E. Rosen, 'Calvin's Attitude towards Copernicus', *Journal of the History of Ideas* 21 (1960), 431-441. Rosen found the quotation in Farrar's Bampton Lectures on the history of biblical exegesis: *History of Interpretation*, London 1886, p. xviii. By the way, Farrar is full of praise for Calvin as an exegete, which makes the inclusion of this miraculous pseudo-quotation even more remarkable.

17 Reijer Hooykaas, 'Science and Reformation', *The Journal of World History* 3 (1956), 136-138.

18 John Calvin, *Institutes* I, 5, 2. Incidentally, this does not imply that Calvin would have advocated an 'independence model' in the sense of section 7.3 in all regards.

19 See Reijer Hooykaas, *Religion and the Rise of Modern Science*, Grand Rapids 1972, 117-122, 154, and compare for the latest 'state of the questions' Davis A. Young, *John Calvin and the Natural World*, Lanham 2007, 43-49.

Nevertheless we find the quotation in which Calvin seems to portray the Copernican world view as trifling with the Holy Spirit repeated in numerous places since White. Even after it was established that the quotation was not to be found in Calvin's work, this still carried on. What is going on here seems to be a textbook example of what is appropriately called *invented tradition*: an artificially constructed tradition, which is carefully nurtured because many people (in this case those who support the conflict thesis) clearly have an interest in it. For example, the quotation figures prominently, with quotation marks but of course without the source ever being given, in various places in the work of Bertrand Russell.[20] And taking his lead from Russell, Paul Cliteur, a leading atheistic voice in the Dutch public realm, has argued even in 2004 that Calvin no less than Luther stood in opposition to the Copernican world view.[21] This 'fact' is part of an argument in which Cliteur supports White's conflict thesis without reservation. It is in this way that well-cherished myths and outdated stereotypes are carefully maintained, though it has of course little to do with responsible historical research.[22]

From this sample of recent research concerning the reception of Darwin, Galileo and Copernicus, we can conclude that the conflict thesis does insufficient justice to the multi-faceted historical reality of the interaction between science and religion.[23] Religion and theology have not always by definition been hostile to scientific discoveries. Conversely, prominent scientists such as Isaac Newton (to give just one of many examples) have often been very devoted believers.

But what about the conflict model in modern times? In response to this question, Barbour points to two movements, which as he sees it understand the relationship between science and religion in terms of a conflict: scientific materialism on the one hand and creationism on the other. Scientific materialism is the view of life according to which the material world (consisting not only of matter but also of energy) is the fundamental reality in the universe, whereas science is the only reliable way of coming to knowledge of that world. More recently it has been

20 For example in his widely read *History of Western Philosophy*, London 1961, 515.

21 Paul B. Cliteur, *De neutrale staat, het bijzonder onderwijs en de multiculturele samenleving*, Leiden 2004, 11, 22. The comment by Luther that Cliteur seems to be referring to comes from his *Table Talks*, and must have been made at a time when it was only rumours about Copernicus' ideas that were doing the rounds (1539); it was only written up in 1566 (so 27 years later!) from memory by one of the guests present. In Luther's authorized work he never refers to Copernicus. See Hooykaas, *Religion*, 121f.

22 On this matter, see also Alister E. McGrath, *Foundations of Dialogue in Science and Religion*, 14-18, who in this regard speaks not entirely incorrectly of an 'urban myth' and 'the perpetuation of outdated stereotypes'.

23 Cf. J.H. Brooke, *Science and Religion. Some Historical Perspectives*, Cambridge 1991, 5ff., 35.

called 'scientism', since the term materialism may misleadingly suggest that according to this view matter is all there is. So clearly, scientism has (like any well-developed view of life) a metaphysical and an epistemological side: it comprises a belief in what reality looks like, as well as a belief in how we can know it. In a more detailed elaboration of this, Mikael Stenmark adds that this scientism also has a rationalistic dimension (it is only rational to believe in what is scientifically justified), together within an axiological one (science is the only useful area of human life) and an existential one (science alone can explain and replace religion, and moreover will solve all or nearly all of our real problems).[24]

Because Stenmark so carefully analyses scientism into its various branches, its ideological character clearly comes to the fore. What we see here is a set of convictions of which it is difficult to see how they can be demonstrated or justified scientifically. It is important to note this, because adherents of scientism often cover up the ideological character of their position, suggesting instead that it is science itself and nothing else that leads to their view. In this way, among many others Richard Dawkins has argued that evolution has shown that the universe does not have a goal. But in doing this, he overplayed his hand, because from the perspective of the natural sciences neither this nor its opposite can be demonstrated. Another well-known practitioner of scientism, Edward Wilson, asserts that any human behaviour, and so also any religious behaviour, can be entirely explained in terms of genetic-evolutionary processes, so that religions have now become obsolete. No doubt they have had useful social functions in the past, but these can quite easily be performed today by science. Like religions did in the past, science is equally well able to unite people around its ideas.[25] But supposing that this the case, why *should* we unite around science? Again, clearly this is not a scientific insight but an ideological belief.

It will be clear that this ideological belief can only relate antithetically to religious belief. After all, according to scientism science and religion focus on the same areas: both are as much concerned with the nature of reality as with the meaning or meaninglessness of it. So whenever science and religion make divergent statements, they immediately come in conflict with one another as a result. And just

24 Stenmark, *Scientism*, Aldershot 2001, 5-14; see also the discussion by T. A. Smedes, *Chaos, Complexity, and God*, Leuven 2004, 6-14, which rightly points out that scientism has become an ideology that is silently at work in Western culture and largely determines our thought forms.

25 See Richard Dawkins, *The Blind Watchmaker*, New York 1987; Edward O. Wilson, *Consilience. The Unity of Knowledge*, New York 1998. For a different account of the significance of the possible evolutionary origins of human religious behaviour, see J. Wentzel van Huyssteen, *Alone in the World? Human Uniqueness in Science and Religion*, Grand Rapids 2006.

as traffic on a major road typically has right of way, so in this sort of conflict science is always right. In such cases, religion only functions as a tiresome road-block to the progress of science.

If we look at this scientistic position using what we learnt in the first half of this book then it will be clear that from the perspective of the philosophy of science, it is closely related to *logical positivism*. There too we found that the conflict thesis was elaborated in a black and white way of thinking about the relationship between science and religion.[26] Here too there was also the idea that the influence of religion on public life was by definition harmful and therefore had to be resisted with the help of science. But we have also seen why this logical positivism quickly turned out to be untenable and was replaced by movements which over time moved further and further away from it. Within scientism all of these later developments in the philosophy of science (from Popper to Feyerabend) are left out of consideration, or at least not fully taken into account. In this sense, scientism (or scientific materialism) must not only be considered as an ideological rather than a scientific position, but also as a position which from the perspective of the philosophy of science is *out-dated.*

If we now look at Barbour's second example of a modern-day movement which is associated with the conflict model then at first glance we have what looks like the exact mirror image of scientism: *creationism*, which takes a literal interpretation of Genesis 1 as its starting point in order to demonstrate scientifically that the earth came into being in six days of 24 hours several thousand years ago.[27]

Creationists seek the same absolute certainty in faith which adherents to scientism seek in science. Both groups are agreed that faith and science deal with the same areas of reality and can thus make competing statements. But, just as according to the latter in such a conflict we must give precedence to our scientific ideas, we should according to the former do the same with our ideas about faith. In response to the rise of Darwinism and of historical biblical criticism, large groups of Protestants, in particular in the United States, often referred to as 'fundamentalists', have propagated this principle just as strongly as the adherents to scientism have propagated theirs (although the latter have usually been able to do this from established academic positions, whilst the former have had to do so more from the side lines). Presumably, both movements are highly dependent upon each other, since both need a clearly defined enemy and find this in one another. In this sense each reminds us of the other time and time again.

26 See for example the autobiographical comments of Rudolf Carnap in Schilpp, *Philosophy of Rudolf Carnap*, La Salle 1963, 7-9.

27 Barbour, 'Models and Relations', 760, talks of 'creation science'; previously, for example in *Religion in an Age of Science*, 8, he used the broader term 'biblical literalism'.

In relation to the conflict model, however, on closer inspection we see a remarkable difference between the adherents of scientism on the one hand and creationists on the other – one that is surprisingly enough only seldom noticed in the literature. For whereas most adherents of scientism express themselves openly against faith and religion, creationists by contrast do not do the same at all in relation to science. They are not opposed in the slightest to the practice of science. On the contrary, due to the fact that they strive vigorously for a scientific foundation to their faith, it seems that they attach great value to science! In this respect, they are even almost a match for the supporters of scientism. Their opposition is only to what in their eyes is bad science, that is to say science that cherishes what they see as incorrect and therefore misleading (for example evolutionary) theories. But in their rejection of bad science they do not of course stand alone, for this is something everyone does …

Strictly speaking then for this reason it is not at all obvious that we should group creationists under the conflict model as Barbour and many others following in his footsteps suggest.[28] After all, creationists strive for the maximum *integration* of science and religion. Believers who make no attempt to support their faith scientifically, because they are convinced that it is right not to ascribe any value to scientific argumentation, can then more justifiably be considered as supporters of the conflict model. They will in fact come into conflict with all those who put great store in scientific research (just as adherents of scientism invariably clash with those who put great store in the cognitive claims of faith). But here we are dealing more for example with the Amish than with the creationists.

To be able to group creationists under the conflict model as well as the supporters of scientism, we must in fact first of all establish that the science they advocate is not a true science but a pseudo-science. For clearly, in that case we can maintain that they are in conflict with all 'true' science. But in Barbour's work this is based on a preliminary choice which remains implicit – and here we come across one of the underlying values of his classification that we pointed to in the introductory section of this chapter. As a matter of fact, given the stubbornness of the demarcation problem that we came across from Popper to Feyerabend, it is by no means easy to distinguish true science and pseudo-science from one another. Moreover, it is questionable whether we should want to do this at all. Feyerabend for one – by no means a religious fundamentalist himself – rightly applauded creationism because it broke through the uniformity of scientific research and in so doing contributed to the methodological pluralism which

28 For example, Michael Peterson et al., *Reason & Religious Belief. An Introduction to the Philosophy of Religion*, New York 1998, 239-242, which very much takes its lead from Barbour's account.

was so desirable in the formation of theories about how the earth came into being.[29]

Nevertheless, creationists are of course in conflict with large parts of established science. Along with Barbour we may ask whether in being so they lose sight of the boundaries between faith and science. In any case, from a Christian point of view there is no need to relate faith and science to each other in the way they do. The approach Calvin pointed to with his doctrine of accommodation does more justice to the specific nature of the Bible and faith in relation to science. For clearly, the Bible does not intend to give us a scientific account of the facts, but communicates with us in the world view of (everyday) language and of ancient Jewish cosmological thought. Given that framework, Genesis 1 firmly declares that God is the creator of the universe, not celestial bodies like the sun or the moon or any other power. This knowledge will suffice for faith, and from this faith perspective the special place and task that belongs to science, e.g. in determining the age of the earth, can be recognized.

This is not to say that the possibility of conflicts between faith and science is by definition ruled out. For clearly, views that arise out of faith (e.g. concerning God's agency in the world) can clash with scientific theories. It is important to note that in case of such clashes it cannot be determined in advance which of both – faith or science – is right. With regard to faith it has regularly been the case that what was considered in the first instance to be biblical and thus correct, on closer inspection turned out instead to be a certain *interpretation* of the Bible (and possibly a wrong one, just as the Vatican has since recognized with respect to Galileo). And when it comes to science, we have seen that the formation of theories is always fallible and provisional; moreover, scientific argumentation is intersubjective, and therefore not strictly compelling. Hence, the history of science is replete with examples of once sacrosanct theories which ultimately turned out to be false. In short, then, when a conflict between science and faith occurs the pros and cons of the conflicting views should be evaluated on a case-by-case basis, since there is no apriori way of determining which one is right.

And wherever absolute infallibility is claimed in the sciences, in for example the evolutionist paradigm in all its various branches, 'foolish' movements such as creationism are very welcome as they break through this pretension. That said, creationism has up to now not really got very far in doing this. It has not been able to develop into an alternative paradigm (Kuhn), or a progressive research-programme (Lakatos). Whether this is also the case for its present-day successor in the academic world, the much more subtle so-called *Intelligent Design* theory according to which we encounter unambiguous traces of design and therefore not only chance

29 Paul Feyerabend, *Against Method*, 30; see also 307.

in the natural world, is still to be seen. Around ID there has been going on a fascinating debate both in the United States and in Europe for some time now, in which emotions sometimes run high.[30] Perhaps, in the end, this is no longer a conflict between faith and science, but rather a debate on scientific methodology. For clearly, the crucial question is what kind of metaphysical ideas can be allowed for: is there, in theory, room for talk of some super-human intelligence in scientific theories, or should we stick to a methodological naturalism according to which no such entities can in principle be assumed? In any case this sort of conflict can be very useful and sometimes lead to surprising new insights which would not be gained without them.

This being so, it will be clear now that the conflict model cannot be seen as an overarching perspective to account for the relationship between faith and science. The relationships between science and religion are much too multifarious and variegated for this to be the case.

7.3 The independence model: a reasonable alternative?

The most effective way of avoiding conflicts between faith and science is to consider them as two separate enterprises that are so different from each other in terms of category that they can in no way whatsoever come into conflict with each other. To give but one simple example: A fish and a bird cannot come into conflict or collide with each another because they each have their own well-defined domain. Similarly, it is easy to understand that many people have tried to improve the difficult relationship between science and religion (and to show the spuriousness of the classical conflicts), by stressing that both have a completely separate domain, task, object and method. So instead of a conflict model they advocate an independence model, which is also known as the 'two-world model'.

Usually the form that this independence takes is such that the sciences have as their task the solution of all questions of fact, whilst faith and religion deal with questions of meaning. As soon as we let these aims overlap, whether it be by allowing faith to answer questions of fact or by giving science authority to speak about questions of meaning, we commit, it is argued, a serious *category mistake*. But if we instead keep them carefully apart, then they can complete each other and science and religion can be understood as complementary enterprises that only together can give a comprehensive answer to our basic questions. Meanwhile, the ways in which this complementarity is fleshed out can differ markedly. Without in any way attempting to deal with this comprehensively, we

30 See for a broad discussion of the main issues Robert B. Stewart (ed.), *Intelligent Design. William A. Dembski & Michael Ruse in Dialogue*, Minneapolis 2007.

shall look in this connection at three very different, but equally influential, movements from the second half of the twentieth century.[31]

In the first place we think of Barthianism in theology. Barth gave an explicit articulation of his vision in the famous *Preface* on the doctrine of creation in his *Church Dogmatics*:

> It will perhaps be asked in criticism why I have not tackled the obvious scientific question [namely that of the doctrine of creation] posed in this context. It was my original belief that this would be necessary, but I later saw that there can be no scientific problems, objections or aids in relation to what Holy Scripture and the Christian Church understand by the divine work of creation. Hence in the central portion of this book a good deal will be said about „naïve" Hebrew „saga," but nothing at all about apologetics and polemics, as might have been expected. The relevant task of dogmatics at this point has been found exclusively in repeating the „saga", and I have found this task far finer and far more rewarding than all the dilettante entanglements in which I might otherwise have found myself. There is free scope for natural science beyond what theology describes as the work of the Creator.[32]

So, with Barth's doctrine of creation we are dealing with something that is strictly theological. Central to it are questions such as what the work of creation says about God and His covenant with His people, how creation relates to salvation in Jesus Christ, and what it means that we human beings are creatures. Each of these are themes that will not interfere so easily with developments in the sciences. In this way Barth manages to avoid the risk of a theology that has to move further and further back as a result of new scientific discoveries. Many scholars have followed Barth in this new definition of the doctrine of creation. We also find that this tendency is very strong in the work of O. Noordmans, who already before Barth but also under his influence had related the doctrine of creation even more rigidly and more exclusively to Christology.[33] At the same time we can ask whether the above quotation is really representative of Barth's position on the relationship between faith and science. In any case, he seems – despite all the space he gives to the sciences, including historical biblical criticism – to attach much theological importance to the *history* that the Bible tells.[34]

31 It is interesting that the independence model has a shorter history than the conflict model (although it does have a long philosophical *pre*-history, from at least the dualisms of Descartes between subject and object and of Kant between nature and freedom). Apparently it is not a framework which would occur to us instinctively whenever we consider the possible relations between science and religion, but it is more a *reaction* to the idea of conflict and to the 'triumphal procession' of the natural sciences.

32 Karl Barth, *The Doctrine of Creation* (Church Dogmatics III.1), Edinburgh 1958, ix-x.

33 O. Noordmans, *Herschepping*, Zeist 1934, 55: 'We shall let the millions of years of astronomy and biology flow calmly by. (...) We shall keep close to the cross'.

34 In this connection, it is interesting that Barth continues the *Preface* of his doctrine of creation by predicting that future authors on the doctrine of creation will no doubt

This is much less the case, however, with Rudolf Bultmann, who can therefore be considered as a more radical representative of the independence model. According to Bultmann, the texts of the Bible are not intended to offer any information about the natural and historical world as it is the object of scientific research. To the extent that the New Testament nevertheless seems to be concerned with earthly reality – for example as the reality in which God becomes human in Jesus Christ, who goes through all sorts of experiences to rise finally from the dead, ascend to heaven and whose return is then anticipated – this is a matter of mythological expression. This is simply what all stories about gods et cetera in antiquity look like. What we have to inquire into is the core meaning of this mythological language, i.e. its meaning for our own existence here and now. For faith is exclusively concerned with this spiritual core meaning that is manifest in the *kerygma* (the message) of the New Testament. So faith does not concern itself with the quasi-historical form into which the *kerygma* is poured. This cannot be so, because our world view is no longer mythic, but scientific in nature. And for this reason according to Bultmann, we can no longer accept the so-called facts of salvation (i.e. the events that Christians remember with Christmas, Eastern, Pentecost etc.) as historical facts, but we ought to examine their particular significance by means of an 'existential analysis' of the New Testament. This particular significance then turns out to be related to our human self-understanding, our choices and decisions and to the call to lead an authentic existence as human-beings.[35]

In Bultmann's work, we therefore see the independence model dealt with much more sharply and clearly than in Barth's. It is clear that Bultmann saw in it a means of 'saving' faith from the jaws of science, after the many conflicts which it had, as he saw it, already lost in this regard. But the price that Bultmann paid for this was that he had to isolate faith from any tangible link with historical earthly reality.[36] Only in this way he could radically leave this historical earthly reality to science. Bultmann made, as it were, a virtue out of a necessity by arguing that the 'objective' earthly reality is not, from the perspective of faith and theology, of critical importance. Much more important is our existential attitude towards it.

come across rewarding tasks in determining the precise borderlines between theology and the sciences (*ibid.*).

35 R. Bultmann, *New Testament and Mythology, and Other Basic Writings*, tr. Schubert M. Ogden, Philadelphia 1984, e.g. 87-89, 107-119; cf. Barbour, *Religion in an Age of Science*, 12.

36 In Barth's work this is also the case, though to a lesser extent; the consequences for him are more limited, inter alia because in the central locus of Christology he places great emphasis on the incarnation, i.e. on the entry of God into the concrete conditions of earthly existence.

On this point, Bultmann was to a large extent indebted to the second twentieth-century movement which strongly supported the two-world model, that of *existentialism* in continental philosophy. Here too we find, as much in the work of theistic thinkers such as Martin Buber as in that of atheistic thinkers such as Jean Paul Sartre and Martin Heidegger, a strong contrast between the sphere of the impersonal objective world and the sphere of the personal world of subjective commitment and engagement.[37] Buber pointed to this difference by distinguishing between impersonal I-It relations and personal I-Thou relations; Sartre distinguished between being-in-itself (*être-en-soi*), that is the bald 'facticity' by which things exist, and being-for-itself (*être-pour-soi*), that is the existence of humans, who thanks to their consciousness do not simply coincide with themselves, but can reflect, make choices et cetera. And Heidegger, with whom Bultmann was in direct contact for several years at Marburg, contrasted the false or inauthentic way of existence, whereby people succumb to all sorts of this-worldly entities, with true or authentic existence, which directs itself towards the true Being. In each case, both worlds are sharply differentiated from one another, and the subjectivity of consciously-accepted human existence is valued much higher than the hard, objective reality which is the object of science and technology.

Of course, when we are dealing with Heidegger and Sartre we cannot talk any longer about a relationship between religious belief and science, because the former plays no significant role in their work. The structure of their thought does however run completely parallel with the two-world model in the discussion about science and religion. As a result, Bultmann was able to appropriate this structure quite easily. For clearly, the common denominator is that the meaning of life is not found in the cold scientific world of hard facts and universal laws of nature, but in a personal commitment arising out of free choices, whereby I as a unique individual – whether or not in God's sight – actualize an authentic existence. That authentic existence does not depend on factual circumstances, let alone scientific developments. Both worlds are completely separate from each other, and for this reason should be carefully differentiated.

The third movement which strongly emphasizes this duality is to be found in Anglo-Saxon analytic philosophy of language. Under the strong influence of Ludwig Wittgenstein, who inaugurated the 'linguistic turn' in philosophy, the focus here was on the different *languages* that science and religion speak. After his force-

37 For an accessible introduction, see Jack Reynolds, *Understanding Existentialism*, Chesham 2006. It is true that Heidegger rejected the term 'existentialism' for his own philosophy, but was in fact an 'existence-philosopher': he belonged to the group of twentieth-century philosophers for whom individual human existence was the central theme.

ful rejection of logical positivism, Wittgenstein became more and more fascinated by the large variety of functions that human language can fulfil. The principle of scientific verification was for him no longer the unassailable norm for a meaningful use of language. Rather, in his later work Wittgenstein considered all sorts of other ways of using language in our lives to be at least as legitimate and meaningful.

A great source of philosophical misunderstanding occurs, however, because we are often not very good at keeping the different ways of using language – the different *language games*, each with its own rules – separate from one another. In such cases, we are 'bewitched' by language. This happens for example when we consider a religious statement as a sort of scientific hypothesis which can be tested empirically. Anyone who thinks this is so has not understood that religious belief constitutes a particular sort of language game, in which a completely different set of rules apply than in the language game of science. Both have their own logic and their own categories. Within the form of life of religion, for example, we have nothing to do with hypotheses, probabilities, proofs and so forth. To think otherwise is something similar to thinking that the equator is a real line around the earth that we can walk over (using an example borrowed from Gilbert Ryle): both ideas are based on a conceptual misunderstanding. In reality, Wittgenstein and his followers assert, the primary purpose of religious statements is not to refer to an external reality. Instead, believers use them to perform completely different linguistic activities. So, for example, they use them to express a certain way of conducting their lives, a certain basic *attitude* to life (for example, one of gratitude, or trust) which they then also commend to others.

Whether this means that religious language does not include any cognitive claims at all is something that is unclear in Wittgenstein's work and is a point of contention amongst his followers. But to the extent that this is the case, such claims (for example that God exists, that God has created the universe etc.) belong to the constitutive presuppositions of faith rather than to its 'core business'. For example, a way of life characterized by thankfulness would be absurd if this thankfulness had no ultimate recipient, just as an attitude to life of trust would be if there were no-one in whom that trust could be placed. However, these cognitive aspects of faith can never be isolated, as if they were individual testable hypotheses. In the context of faith they cannot be called into question, rejected or tested. But the language game of faith *as a whole* can at any given moment be in danger of losing its function, as a result of changed cultural circumstances. The sciences can of course play a role in such a change in cultural circumstances, so that it is in fact possible that in an indirect way faith and science can come into conflict.[38]

38 For this interpretation of Wittgenstein, see for example: Vincent Brümmer, 'A Dialogue of Language Games', in: id. (ed.), *Interpreting the Universe as Creation*, Kampen

In the extreme version of Wittgensteinianism which rejects the idea that religions have any cognitive aspects at all – a version that is sometimes termed 'Wittgensteinian fideism'[39] – we see the independence model coming to the fore in a purer form. For here the possibility a conflict between faith and science is excluded beforehand. Science is concerned with the explanation of phenomena in reality, religion with the articulation of the meaning and purpose of reality. These are two very different things. Anyone who loses sight of this disparity between them will therefore be guilty of 'smothering the differences' as Gilbert Ryle called it.[40] Or, to use another term of Ryle's which has become famous: he or she commits a *category mistake*, like the doctor who with reference to Romans 6:23 gave the cause of death for one of his patients as 'sin'.

The two-world model has been particularly popular for some time now. And new defences of it continue to appear, in particular in situations where people either set themselves up against what they see as misguided battles by religious believers, and/or want to oppose any 'contamination' of faith with science by adherents of scientism. We meet an example of the former in one of Stephen Gould's popular books, in which he strongly defends the principle that he calls NOMa. NOMa stands for 'Non Overlapping Magisteria', where a *magisterium* (from the Latin *magister* = teacher) is a kind of teaching authority. Gould asserts that the teaching authority of science is concerned with how the empirical universe fits together, whereas the teaching authority of religions is related to moral values and the meaning of life. Because these are two different categories, there is always clear blue water between them, which is contrary to what both creationists and adherents to scientism believe. Science deals with the 'the age of rocks', whilst faith is concerned with 'the rock of ages'.[41]

An example of the latter – people defending the independence model in opposition to the confusion of science and religious faith by adherents of scientism – we find in a recent book of Taede Smedes. Here, he uses the biblical account of Balaam who beats his ass to make it stand up and carry on walking (Numbers 23) as a metaphor for the present state of affairs in the religion and science debate. Just as Balaam tried to impose his will on the ass, so too does scientism attempt to force its standard of rationality on theology, by for example forcing it to provide

1991, 1-17. A more extreme interpretation of Wittgenstein is to be found in, amongst others, many publications of D. Z. Phillips.

39 Since Kai Nielsen, 'Wittgensteinian fideism', *Philosophy* 42 (1967), 191-209. D. Z. Phillips, though, has always rejected this phrase as being an inappropriate description of his own position.

40 Gilbert Ryle, *Dilemmas*, Cambridge 1954, 80-81; cf. Brümmer, 'Dialogue', 1-2.

41 Stephen Jay Gould, *Rocks of Ages. Science and Religion in the Fullness of Life*, London 1999.

insight into the activity of God in the world by means of modern scientific theories. Theologians who accept this challenge commit, according to Smedes, a category mistake, for science can say absolutely nothing about God: the activity of God can only be perceived with the eyes of faith, for it does not breach the natural order.[42]

It can be very useful indeed to emphasize the structural differences between faith and science whenever there is a danger of their boundaries being transgressed (either in a synthetic or in an antithetic direction).[43] In this sense the independence model is certainly of importance. Nevertheless, in the rigid form in which it is often presented, it is unconvincing. During our review of the present state of affairs in the philosophy of science, we already saw that the Theory of the Neat Dividing-line must be considered outdated. The dividing-lines between science and religion are at the very least not water-tight. Anyone who denies that religion and faith have any empirical significance or cognitive content does not do justice to the way in which the vast majority of believers conceive and experience their faith. Likewise it is also undeniable that at least *some* scientific theories and paradigms tend to have ideological and moral implications. Can for example someone who, on scientific grounds, thinks that the moral values that we hold to are purely a function of the struggle for survival by our genes at the same time still ascribe objective validity to these values? Perhaps this is possible, but in any case it is not easy to see how this can be so. For if norms and values can satisfactorily be explained as useful survival mechanisms of our genes, why should we then have to hold to them as individual human beings? Why should we support the struggle of our genes?

More generally we now note briefly the following objections to the independence model.

1. As we saw already in chapter 1, Western science has an aura of all-comprehensiveness about it. It does not allow itself to be limited to just a part of what is important in life, but rather has the tendency to consider that which falls outside its conceptual frameworks – what cannot be empirically observed – as unreal or at least unimportant. This tendency is not to be ascribed to individual scientists, but in one way or another it is part and parcel of the scientific project as such. Anyone who is involved in the neuro-sciences, for example, can hardly avoid the impression that humans are little more than robots guided by their brains. We do not come across such things as freedom and self-consciousness

42 Smedes, *Chaos, Complexity, and God*, 157-158 (Smedes's book started life as an unpublished doctoral thesis, Groningen 2004, the title of which was: *Avoiding Balaam's Mistake*).

43 See e.g. Arjan Markus, *Beyond Finitude. God's Transcendence and the Meaning of Life*, Frankfurt 2004, 370-378, where he carefully disentangles the physical and metaphysical claims which are incorrectly confused with each other in the science and religion debate.

here. So there is an inherent tendency in science to interfere with issues that pertain to our view of life and our personal existence – a tendency which is denied or at least made invisible by the independence model.

2. In a similar way Christian faith – and again we shall limit ourselves to this over against abstractions such as 'religion' – cannot tolerate it when its relevance is limited to only a part of reality or of what is important in life. This is the consequence of the Christian *doctrine of creation*. For clearly, according to this doctrine God does not merely have a relationship with our spiritual existence and our moral values, but with 'heaven and earth', including the concrete materiality of the earth. His reign is not only related to the realm of grace, but covers nature and history too – and so we see something of this from time to time in nature and history. The alternative view, according to which God must only be seen as relating to a spiritual salvation-event and not to the concrete reality of earthly existence was left behind by the Christian church very early in its history when it rejected Marcionitism.

3. The believing practitioner of science is at risk of getting a split personality. For clearly, such a person has to live in two totally different worlds which can hardly be related to each other. So he or she cannot escape a great discrepancy between 'on Sundays' and 'during the week'. It is of course true that both realities can be considered complementary, but only in the way that iron and clay are complementary: they do not mix well.[44] Such a life-in-two-worlds can only be lived in a more or less artificial way, which stops people from becoming fully-integrated human beings.

4. The net result of the independence model has been a huge loss of relevance for faith. This has come about not only because the all-encompassing claims of science speak more to the collective imagination than the universal claims of faith. But it has come about above all because Christian faith simply hangs in mid-air when it is not allowed to relate any more to the concrete, earthly, empirically-perceptible reality.[45] In general, religions (apart, perhaps, from gnostic belief systems) cannot be cut loose from empirical reality without their identity being seriously affected. In this sense, the independence model easily becomes a secularizing force.

5. As we saw in our survey of the philosophy of science (in particular in chapter 4) there is no watertight division between science and religion, because

44 Cf. the Book of Daniel 2:33, 42.

45 Wolfhart Pannenberg, *Systematic Theology* 1, 232, is right to say that it is 'tempting' (German: *verführerisch*) to avoid the question of the historical reality of the biblical tradition with respect to God's salvific acts, or to deny its importance. The reality of faith and the 'soberness and seriousness of belief in the God of the Bible' depend on it.

science necessarily arises from underlying convictions that cannot be justified scientifically. Thus, as Polanyi and others have stressed, in science we have to do by definition with matters of faith and trust. And even though this faith is not per se religious in nature, it still is a deeply rooted non-rational perspective from which reality is viewed. Conversely, religious as well as non-religious views of life always furnish us with certain background beliefs which play a role in scientific research, for example in the area of theory choice.

For these reasons, we can conclude that the independence model, despite the good and useful aspects that it has, falls short as a comprehensive perspective with which to articulate the relationship between faith and science. This is also the case for the currently most popular variety of it, the so-called *methodological naturalism* or *methodological atheism*. Here, in general terms, the idea is that in scientific research no appeal may be made to supernatural agents – not because they do not exist, but because it would be methodologically incorrect to do so. Also bringing in other beliefs which stem from someone's faith would be incorrect. Religious convictions ought not to play any role at all in scientific research.

Now in general it is indeed advisable when looking for scientific explanations not to refer too quickly to acts of God or other transcendent beings. For clearly, this kind of move can easily function as a 'science stopper', whilst further study often brings to light that a satisfactory natural explanation of the phenomenon in question does exist. Nevertheless, on the whole we cannot say categorically that in scientific research, 'God must be kept away'. Firstly, this view is based on the preconception that a certain observation (for example a miraculous healing) cannot be adequately explained as the result of a direct intervention of God. But how do we know that such divine interventions are in fact impossible? The fact that in the past people often jumped to this sort of conclusion too quickly does not imply that it is therefore wrong at all times and in all places. It is much better not to exclude certain options a priori in scientific research, but rather to be open in principle to all possibilities.

Secondly, we have already seen that it is very doubtful whether it is at all possible to prevent religious or other ideological and metaphysical convictions from playing any role at all in scientific research. Even methodological naturalism is a 'metaphysical' position, which must of course be adopted in order to practice science in the way it requires. Moreover, we remember the under-determined nature of theories by observation and the connected questions concerning theory-selection. It is much more rational when we are dealing with theory-selection to bring into play everything that we think we know rather than simply that part of it that is not related to our view of life. If for example someone is convinced, on religious grounds, that humans are not robots, but rather responsible and in a certain sense also free beings, and the results of the brain research that they do can be explained just as well on the basis of a theory which is in line with that conviction as on another, may they then not allow their religious convictions to decide which theory they choose? It is not clear why this should not be so.[46]

In conclusion, we shall have to look for a third model which will hopefully be able to bring more clarity to the question of how and where the domains of science and religion relate to one another and can legitimately influence each other.

7.4 The dialogue model as theological challenge

During the past fifty years or so, a growing number of academics have tried to overcome the strict dichotomy between science and religion that was proclaimed by adherents of the two worlds model. In the Netherlands, for example, the chemist C.J. Dippel and theologian J.M. de Jong – pupils of Karl Barth and Rudolf Bultmann respectively – took the lead in attempting to map a way forward from the independence model to the dialogue model already in the middle of the former century.[47] Others were to follow this path with an increasing degree of confidence. In the slipstream of the developments in the philosophy of science which we sketched earlier in this book, it seems that even the majority of those who profesionally study the relationship between science and religion are exploring the possibilities of an ongoing *dialogue* between both. With a term introduced by Ernan McMullin, they are looking for what is called a new *consonance* between them.[48] Conflict- and independence models have been left behind, and, inspired by the conviction that both

46 For the two arguments we have discussed and for some others against methodological naturalism, see Alvin Plantinga, 'Methodological Naturalism?', in: Jitse M. van der Meer (ed.), *Facets of Faith and Science*, Lanham 1996, part I, 177-221 (the four parts of this publication happen to form a veritable treasure trove of material about the interaction between faith and science). Various versions of this article can also be found on the internet.

47 Cf. C.J. Dippel & J.M. de Jong (eds.), *Geloof en natuurwetenschap. Studies over de verhouding van christelijk geloof en moderne natuurwetenschap*, The Hague 1965 (part 1) and 1967 (part 2). As to the situation in Germany, the influential work of the Christian physicist and philosopher C.F. von Weizsäcker comes to mind, e.g. among many other works his *The World View of Physics*, London 1952 (German original: Leipzig 1943, with many later reprints) and *Die Tragweite der Wissenschaft*, Stuttgart 1964 (6th enlarged edition 1990).

48 Ernan MacMullin, 'How Should Cosmology Relate to Theology?', in: Arthur Peacocke (ed.), *The Sciences and Theology in the Twentieth Century*, Notre Dame 1981, 39. Cf. Ted Peters (ed.), *Science and Theology. The New Consonance*, Boulder 1998. The new journal *Theology and Science* explicitly asserts that its goal is to build bridges between faith and science and to uncover the consonance between them (without however forgetting the dissonances between them). See the editorial articles by Peters and R. J. Russell in the opening issue, *Theology and Science* 1 (2003), 1-8.

can benefit from this, people are trying to bring faith and science into conversation with each other in a constructive and responsible manner. This happens on a whole range of fronts and in a whole variety of ways, of which we can only mention a few briefly in passing. What is important here is that in the dialogue model one tries to respect the specific nature and significance of both faith and science. The metaphor of consonance expresses this wonderfully: when two voices or instruments are consonant (literally: sound together) it is not the case that each makes the same sound, but rather that the *different* sounds they produce make an unmistakeable harmony. The harmony between faith and science that people attempt to discover in the dialogue model likewise does not so much have the character of overlap, but rather of parallels and analogies and of tendencies which point in each other's direction.

There is good reason to seek such consonance because, as will have become clear, the modern practice of science has important roots in the Judaeo-Christian faith traditions. To state it more precisely: important conditions which have enabled the rise of Western science are to be found in the Judaeo-Christian belief in creation. Amongst these conditions are the belief that the material reality is real, that it is profane and not sacred, that it is valuable, that it is contingent, and that it manifests a certain order which can in principle be apprehended by human understanding. At first sight each of these seems to be a banal statement of the obvious. But from the perspective of world history this is certainly not the case. It can of course hardly be coincidental that the very successful practice of science as we know it happened to develop in Europe of all places, and precisely at the point at which the influence of Greek philosophical ideas (especially those of Aristotelian descent) began to wane there during the sixteenth century.

If for example we were to consider the material world not as a real but as only existing in appearance and therefore deceptive, as is the case in some Eastern religions, then we would not be inclined to study it intensively. If we saw it as sacred and permeated with all sorts of divine powers, as in animist religions, then we would hardly *dare* even to study it. If we did not consider reality to be valuable, but rather as less valuable than the ideal world, as in Platonism, then this too would reduce the likelihood that we would make an intensive study of it. If we considered material reality not to be contingent, but rather necessary, as in Aristotelian thought, then we would think that we could know its structure for sure from certain fixed principles (such as the fact that the planetary orbits must always be round), and that further study would therefore be superfluous. On this basis, Aristotle could believe that women had less teeth than men, and did not find it necessary to check this by actually looking at a woman's set of teeth. In general the Greeks were stronger in logic and mathematics than in experimental science.[49] If, finally, we believed that material reality were, to be sure, contingent, but at the same time not rationally ordered and therefore impossible for human understanding to comprehend, we would only find chaos in the natural world.

It is precisely in the Judaeo-Christian doctrine of creation that the material reality is seen as real instead of apparent, as profane instead of divine, as valuable instead of negligible, as contingent instead of necessary, and as rationally ordered instead of structurally chaotic.[50] After all, God created the world out of nothing as a non-divine reality outside Godself. This act of creation is a sign of God's goodness, so that the cosmos can be none other than valuable. Given that God is free in his actions, we cannot then establish a priori from our own thought which laws and structures creation is subject to: creation is contingent, that is, it could have been different from how it actually is.[51] We therefore have to study it in order to discover how it in fact works. Moreover, this is not in vain given the fact that humans are made in the image of their Creator – an idea that was traditionally closely connected to the faculty of *reason* as common to God and humanity. Because God went about creating the world in a rational manner, despite the pervasive influence of sin it is still possible for us human beings to understand the world with the aid of reason. Taken by itself this is far from self-evident. As Einstein is reported to have said: the most incomprehensible thing about the universe is that it is comprehensible...

Now, we must of course not over-estimate the significance of this sort of Christian basis for the modern practice of science. For outside the *corpus Christianum* we also find experimental research which can be seen as the precursor to the methodical practice of science, as in the Arab world during the Middle Ages. Furthermore, all sorts of other factors played a role during the time of the Renaissance in Europe in the rise of modern science: the increasing trade and industry, humanism with its emphasis on education and scholarship, et cetera.

Moreover, the fact that science does to a large extent go back to Judaeo-Christian roots should not only be considered as something positive. For clearly, as we briefly mentioned in chapter 1, science has not only given us much that is good, but also much that is bad. We only have to think of the exploitation of

49 Barbour, *Religion in an Age of Science*, 17.

50 Apart from Barbour, see on this point Thomas F. Torrance, *Divine and Contingent Order*, Oxford 1981. Alister E. McGrath, *Foundations*, 53-55 also points to the *uniformity* of nature as an unprovable but necessary assumption in science, which seems to be inspired by the Christian belief in the reliability of God as Creator.

51 For a sophisticated historical and systematic reconstruction of this notion of contingency, which shows its relevance for an adequate ontology, see Antonie Vos et al, *John Duns Scotus, Contingency and Freedom*, Dordrecht 1994.

nature, the pollution of the earth and other huge environmental problems, the development of weapons of mass destruction and so on. The Judaeo-Christian doctrine of creation led to a de-sacralization of the way we deal with the natural world and thus an increasingly matter-of-fact approach to that world. One result of this was that it legitimised the shift towards an unbridled exploitation of nature.[52] Although we can rightly ask here whether misunderstood or even misused versions of the doctrine of creation in question are at work here, the ecological consequences are nevertheless no less fundamental.

So, in both a positive and a negative way the Judaeo-Christian doctrine of creation has been an important driving-force in and source of stimulation for the development of science as we now know it.[53] When we come to the question of whether this development would have been possible from other religious backgrounds, we can only speculate. It is interesting that for example Chinese astronomy was at a more advanced stage at the time of Galileo than its Western counterpart: it had confirmed the appearance of a great number of stars, and even managed to observe sunspots without the aid of a telescope. Nevertheless, it was only the developments in Europe that led to what is usually called the scientific revolution.[54]

And here in practice in the seventeenth century, it were often deeply-convinced Christians, of all people, (amongst whom were many theologians) in the frontline of promoting experimental research and other forms of scientific practice. For them it was a way of getting to know the Creator better and to marvel at his power and wisdom. Article 2 of the *Belgic Confession*, which became authoritative among Reformed Christians soon after its publication in 1561, stated that 'we know God first of all by the creation, conservation and government of the whole world, for this is before our eyes as a beautiful book …'. Clearly, the old metaphor of nature as a *book* that can be interpreted was taken up here in a way that stimulated the empirical study of nature.[55] To the question which is

52 This insight has been gaining ground especially since Lynn White's essay 'On the Historical Roots of our Ecological Crisis' in 1967, which was republished in Roger S. Gottlieb, *This Sacred Earth. Religion, Nature and Environment*, New York 1996, 184-193.

53 For this point, also see Hooykaas, *Religion and the Rise of Modern Science.*

54 Although it is therefore not possible to say that 'only Christianity fostered science', it seems to me to be a misplaced form of political correctness to portray this view of things as 'deeply misleading' and to speak instead of a 'profound and longstanding engagement between science and religion in all literate cultures', as Wentzel van Huyssteen does ('Introduction', *Encyclopaedia of Science and Religion*, xi.)

55 See on the background, reception and continuing relevance of this article my inaugural lecture *Als een schoon boek*, Leiden 2008. For the *Belgic Confession* as a whole, see Nicolaas H. Gootjes, *The Belgic Confession. Its History and Sources*, Grand Rapids 2007. For the pedigree of the 'book of nature' metaphor, see Arjo Vanderjagt & Klaas

sometimes asked of whether believing scientists should not aim for something like a 'Christian science' we should perhaps answer that in fact science *is* a Christian enterprise, deeply anchored as it is in the doctrine of creation! The fact that later generations of scientists distanced themselves from these Christian roots, just as a booster is discarded from a rocket once it has helped to launch it, is another matter. The consonance between Christian faith and the presuppositions on which science was built does not change as a result of this development. As a consequence, we can understand why, after the periods in which the relationship between faith and science was seen above all in terms of conflict and then total independence, there is now a search once more for this consonance: from a historical point of view, there *is* a strong link between both. We see this not only when it comes to the presuppositions on which science is built, but also when we consider important methodical parallels and the mutual contributions made by science and religion to so-called boundary issues.

Regarding the *methodical parallels* between science and religion we will not discuss again what we already saw in our overview of the newer philosophy of science in the first part of this book: the important role of personal commitment in both areas, the function of traditions which people gradually grow into, of over-arching paradigms, the fact that both sensory and religious experiences are always interpreted in interaction with some pre-existing perspective, et cetera.[56] Instead, we shall now focus on two further methodical parallels: the adoption or assumption of critical realism in both science and religion, and closely connected to it, the role that models and analogies play in both fields.

The term *critical realism* came up in the discussion on the question of whether scientific theories and especially unobservable entities that are postulated in them literally represent reality in, as it were, a one-to-one relationship, or whether they are simply useful instruments that we devise on our own in order to help us understand what happens in reality. In the first case we talk of 'realism', in the second of 'instrumentalism'. The question is not so much about concrete objects that everyone can observe (here everyone is more or less a realist), but more about that part of reality that cannot be observed with the naked eye, such as that which is the object of micro- and macro-physics. If we turn this into a concrete question, it would be whether for example electrons 'really exist', or whether the notion of an electron is a conceptual tool that scientists have constructed to be

van Berkel (eds.), *The Book of Nature in Antiquity and the Middle Ages*, Leuven 2005. For its influence on the emerging science, cf. Kenneth J. Howell, *God's Two Books. Copernican Cosmology and Biblical Interpretation in Early Modern Science*, Notre Dame 2002.

56 See Barbour, *Religion*, 22, who furthermore, following Polanyi, also points to the pursuit of rationality and clarification in both areas.

able to correctly predict certain perceptible phenomena. In that case, electrons do not exist in reality but are only useful fictions in the minds of researchers.[57] Historically, it was the Protestant theologian Andreas Osiander (1498-1552) who already argued in a similar way that the Copernican system was a hypothesis which served as a useful basis for astronomical calculations, but which did not necessarily have to be true, in the sense of reality-depicting.

The dilemma of scientific realism versus instrumentalism (or anti-realism) has haunted the philosophy of science from its very beginning.[58] Logical positivists were divided over the issue, with Schlick taking a realist stance and Carnap sympathizing with a variety of anti-realism. According to the latter, the acceptance of theoretical frameworks within which unobservable objects are postulated 'do[es] not mean for us anything more than acceptance of ... new linguistic forms.' Phrases about new entities 'must not be interpreted as referring to an assumption, belief or assertion of „the reality of the entities"'.[59] In more recent times, a sophisticated form of this position has been vigorously defended in a series of publications by Bas van Fraassen. Van Fraassen argues that theoretical terms that refer to unobservable entities are introduced into scientific discourse for pragmatic reasons, not on epistemic grounds. They improve the reliability of empirical predictions, but since we cannot observe them, we ought to be agnostic about whether or not they exist. Strictly speaking, Van Fraassen is not an instrumentalist, since he does not deny that unobservables exist but only advocates agnosticism here: given the empirical data, we simply cannot know whether or not electrons et cetera exist.[60] Van Fraassen is contradicted, however, by contemporary realists such as Ernan McMullin (whom we already mentioned) and Hilary Putnam, who argue that by far the most adequate – if not the only sound – explanation for the empirical success of scientific theories, is that these theories truthfully depict the unobservable parts of the world.

Now in the field of faith and theology a similar discussion is currently going on concerning non-observable entities which are important here, in particular of course the reality of *God*. Realists think that in religious language statements

57 For nice summaries of the issues, see Okasha, *Philosophy of Science*, 58-76; Christopher Hitchcock (ed.), *Contemporary Debates in the Philosophy of Science*, Oxford 2004, 115-148; Dooremalen, De Regt & Schouten, *Exploring Humans*, 325-360.

58 Van Dooremalen, De Regt & Schouten, *Exploring Humans*, 326-329.

59 R. Carnap, *Meaning and Necessity*, Chicago 1956, 31.

60 Van Fraassen himself labels this position 'constructive empiricism', since it represents science as engaged with theoretical constructions on the basis of empirical (i.e. observable) data. See his seminal work *The Scientific Image*, Oxford 1980, 5. A later elaboration of his views can be found in Van Fraassen, *The Empirical Stance*, New Haven 2002.

about God are to be taken literally, which means that they intend to describe the divine reality. Anti-realists, on the other hand, like the (extreme) Wittgensteinians whom we met in the previous section, are inclined to say that religious belief claims can never be separated from the framework or form of life to which they belong.[61] According to them such claims are misinterpreted when they are taken as referring to a supposed absolute reality existing separate from any linguistic framework. In this way, people are then often misled by the apparent descriptive character that religious language often has, whereas in fact it fulfills quite other functions. Wittgenstein at one point tried to make clear the special character of religious language by looking at the expression 'the eye of God'. When talking of the eye of God that sees everything, believers usually have in mind that God knows all that happens, both in the world and in the hearts of human beings. But then Wittgenstein rhetorically asks: 'Are eyebrows going to be talked of, in connection with the Eye of God?'[62] The expected answer is: Of course not. For clearly if we asked after God's eyebrows in this connection, we would betray that we had overlooked the fact that the expression in question is used in an instrumental, metaphorical way rather than literally. Now this is not only the case for certain individual expressions but for religious language as a whole. According to some who have thought further along these lines we only speak of God in order to prompt ourselves towards a morally serious way of living or to an ultimate confidence, or towards seeing meaning and significance in our lives. In such cases there would be no extra-linguistic reality that corresponded to religious statements about God, but they would have other useful functions that could not easily be taken on by other concepts.

The physicist and theologian John Polkinghorne has pointed out that most scientists as well as most believers tend towards an intermediate position between these extremes. This intermediate position is often referred to as 'critical realism' (and then the classical form of realism is often considered as 'naïve realism' to make a clear distinction between the two) or as 'moderate realism'.[63] In contrast to instrumentalism and constructive empiricism, critical realism asserts that entities which are not directly observable may nevertheless exist and can

61 There is a striking similarity here with Carnap's view of scientific nonobservables, as developed in *Meaning and Necessity*, 20-40; see the quotation given above (and note the reference to a 'framework' here).

62 Wittgenstein, *Lectures and Conversations on Aesthetics, Psychology and Religious Belief*, ed. C. Barrett, Oxford 1966, 71.

63 For Polkinghorne's own variety of critical realism see e.g. his *Belief in God in an Age of Science*, New Haven 1988, 104 ff.; cf. Smedes, *Chaos, Complexity, and God*, 33-39. Other well-known critical realists in the area of science and religion are Ian Barbour and Arthur Peacocke.

also be known. In contrast to naïve realism, however, critical realism asserts that our knowledge of reality (including its non-observable parts) is always *mediated*, that we as knowing subjects play a co-determining role by means of our interpretative schemes, and that our knowledge is therefore never complete and objective but always analogical and partial. So our knowledge of reality is not based on a one-to-one representation of it. On the contrary, we remain continuously engaged in articulating our experiences in such a way as to do maximal justice to this reality. So we are truly in touch with reality,[64] but because we have no direct access to it (especially to its non-observable parts), what we say about it continues to be tentative to some extent.

In fact many believers will justify what they say about God in this way. They will admit that no-one has ever seen God. So, how then can people claim to know God? This knowledge can only be mediated knowledge. For Christians this is not a concession which they have to make willy-nilly, but a confession: it is the basic conviction of Christianity that the knowledge of God is *mediated* by the human being Jesus Christ. As the only-begotten Son of God Christ has made God known to us (John 1:17). As all knowledge, this knowledge as well 'connects' us to a reality that does exist, not to some fictitious pseudo-reality. In that sense it is trustworthy. At the same time, however, it can never be isolated from a personal relationship with God characterized by faith; it can never be objectified, or broken down into x number of propositions. We can only speak of the reality of God in concepts and images which approximate it, without ever fully grasping it. In this sense it is true to say that our knowledge is 'in part' and that we see 'in a mirror (darkly)' (I Corinthians 13:12). It is not the sort of knowledge that mathematics provides us with, but it is typically 'experiential' knowledge that is nourished by the experiences of faith.

A second methodological parallel between faith and science is to be found in the significant role of *models*, *metaphors* and *analogies* in both fields. In fact, this parallel is closely related to what we have just discussed. For clearly, if the realities that science as well as faith deal with cannot be known directly or described exactly, then it is obvious that we need a special language to get as close to them as possible. And this is what we have. In the Bible it is already recognized how important images and parables are in talking faithfully about God. For images (or metaphors) and parables indicate partial correspondences without specifying precisely what these correspondences are. As a result they are pre-eminently suited to pointing to a reality that transcends us and fills us with wonder and

64 Roger Trigg, *Rationality and Religion*, Oxford 1998, 83-89, especially stresses this aspect of critical realism: it allows us to see our subjectivity as *mediating* our contact with reality, rather than as impeding it.

amazement. A metaphor such as 'the Lord is my shepherd' for example performs just such a function. No 'literally true' statement is intended by it, for God is of course different from a shepherd in all sorts of ways. But what is said with it is that God is like a shepherd for the speaker *in certain regards*. Sometimes, such a metaphor is developed further, as in Psalm 23, or supplemented with or limited by other images. In the New Testament, the parables that Jesus used in his teaching can be seen as elaborate images of this kind (sometimes worked out in great detail). They do not give us any precise definitions or exact descriptions of God and His Kingdom, but do indirectly bring light to bear on certain important aspects of them.

Now when it comes to physics we might think that we are dealing with a much 'harder' reality, which we can in fact talk about more straightforwardly in precise concepts and definitions. Niels Bohr, however, considered quantum theory, of which he was one of the founders, as a good example of a set of facts that had been understood in all clarity but about which we can only still talk in pictures and parables:

> Images and parables are concepts like wave and particle. They do not precisely reflect the real world, but they do so in part. This is probably also the case with the problems of metaphysics. (...) If the religions throughout the ages have spoken in images, parables and paradoxes, then this can only mean that there is no other way to conceive of the reality that they are used to refer to. But this does not mean that there cannot therefore be any such reality.[65]

And it is not only in quantum mechanics, but also in other areas of natural science as is well-known a lot of work is done with analogies and models in order to describe complex phenomena adequately.[66]

Now such parallels do not of course 'prove' anything, all the more so since we can point to important differences between the use of models and other images in science and religion.[67] The dialogue between science and religion be-

65 This quotation goes back to a compilation of statements ascribed to Bohr in W. Heisenberg, *Der Teil und das Ganze. Gespräche im Umkreis der Atomphysik*, Munich 1969, 285, 123. No doubt these statements are also included in the English translation: Heisenberg, *Physics and Beyond. Encounters and Conversations*, New York 1971; but I haven't managed to get hold of a copy of this translation.

66 For the use of models in science and religion, see also Pieter Huiser, *Models, Theories, Narratives. Conditions for the Justification of a Religious Realism*, Amsterdam 1997.

67 Alister McGrath, *Science and Religion. An Introduction*, 144-146. No doubt, one of the more important differences is that in the natural sciences models can successfully be used to predict certain phenonema, whereas in theology, as well as in the social sciences and the humanities, this is much more difficult, if not totally impossible. In that sense, the natural sciences are 'harder' than e.g. theology. (I am grateful to the physicist Remko Muis for pointing this out to me).

comes more substantial, however, when we look at those issues where they clearly tend towards each other, and, as it were, point in each other's direction. This seems to be the case in the mutual contributions that both science and religion make in the debate on so-called *boundary questions*, that is questions on the boundary of what we can study scientifically. And here we are dealing in particular with the microscopically small and the astronomically large cosmos. In the following paragraphs, we will just mention the most notable substantial issues which are being debated in this connection, without being able to spell them out extensively.

Regarding the former – that is, the subatomic level of reality – we may think again of the development of *quantum mechanics* in the course of the twentieth century. Apparently, this required us to reconsider the role of the observer or the knowing subject in scientific research. For it suggested that our observation of quantum phenomena is unavoidably influenced by the presence and involvement of the observer in an experiment. More than anywhere else the presumed 'neat dividing-line' between subject and object seemed to be porous in quamtum mechanics. As a result, our knowledge of quantum phenomena is always indeterminate and incomplete: if we increase the precision in measuring the *position* of a quantum particle, we are forced to lose precision in measuring its *momentum*, and vice versa. However, Werner Heisenberg has convinced many that the situation is even more complicated. For according to his famous 'uncertainty principle', the incompleteness of our knowledge in this regard is not simply a special instance of the 'observer effect', which might disappear as soon as we find a means to circumvent the influence the observer exerts on the observed phenomena. Rather, states with both definite position and momentum *just do not exist* in the quantum world. So the uncertainty is not due to some imperfection in our measurement equipment, but is an irreducible ontological feature of the sub-atomic reality it-self – at least according to this so-called Copenhagen interpretation (Copenhagen being the place where Heisenberg, together with Niels Bohr, worked at that time). As a result, it must be concluded that the quantum world is *indeterministic*, that is to say it is not bound up in a closed system of cause and effect.

This is an idea that is so contrary to classical physics that the opposition to it and the divisions arising from it within the world of natural science are quite understandable. Albert Einstein, for example, consistently rejected the indeterministic view ('God does not play dice'). But although there are still viable alternatives, such as David Bohm's 'hidden variables' theory, at present the indeterministic interpretation is generally preferred by physicists and philosophers of physics.

Now such indeterminism might have interesting consequences for the dialogue between science and religion. In particular, it would mean that the way in which a divine intervention in the world was ruled out by science in the past, with an appeal to closed causal relationships which were said to determine all natural processes from beginning to end, would now no longer be convincing. In this connection, some have argued that the inherent flexibility and openness of the quantum realm provides room for God to act in the world in a way that does not violate the natural order. Others, however, have countered this suggestion, for example by arguing that it is not easy to see how the peculiarities and uncertainties at the quantum realm may affect the much more ordered higher level of everyday life. In this way, a leading figure in the science and religion debate such as John Polkinghorne has voiced reservations over this form of theological 'exploitation' of the Heisenberg principle.[68]

According to Polkinghorne, the so-called *chaos theory* is much more promising in this connection for the dialogue between science and faith. Chaos theory explains why it is, for example, still not always possible for us to predict tomorrow's weather with certainty. The processes which determine the weather are in fact chaotic. This means that miniscule differences in initial conditions can lead to enormous changes in the weather. To make this clearer, the example is often used of the so-called butterfly effect: a little movement of air, caused by a butterfly flapping its wings, can set in motion a whole series of events which will lead five days later to a tornado in a completely different part of the world. According to Polkinghorne, chaos theory pre-supposes an indeterministic universe and it therefore provides sufficient conceptual space to allow for an action of God, by influencing chaotic systems, to make a difference in the universe. Many physicists however do not agree with Polkinghorne that chaos theory allows for an indeterministic interpretation.[69]

Natural science tends, even more emphatically, in the direction of a renewed dialogue with faith on the matter of *cosmological* boundary questions. Here, the dialogue has as its main focal points the *Big Bang theory* and the so-called anthropic principle. The Big Bang theory is seen by some as providing great support for the Judaeo-Christian doctrine of creation *ex nihilo*. For clearly, according to this theory the cosmos has not always existed, but came into being out of nothing at a certain moment. William Craig, who has occupied himself a great deal with

68 John Polkinghorne, *Science and Providence*, London 1989, 27-28. For an overview and critical discussion of proponents of a 'quantum theology of divine action', see N.T. Saunders, *Divine Action and Modern Science*, Cambridge 2002, 94-172. Quantum physics also played an important role in the series of conferences on divine action, held between 1992 and 2000 and sponsored by the Vatican Observatory and the Center for Theology and the Natural Sciences. See esp. Robert John Russell et al. (eds.), *Quantum Mechanics. Scientific Perspectives on Divine Action*, Vatican City State 2001.

69 See on the relationship between chaos theory and theology another volume in the 'divine action-series': Robert John Russell et al. (eds.), *Chaos and Complexity. Scientific Perspectives on Divine Action*, Vatican City State 1995.

cosmological questions, even speaks in this regard of a 'dramatic empirical verification' of the biblical belief in creation – which seems to be somewhat of an exaggeration.[70]

In its strong form, the *anthropic principle* asserts that the possibility of life – and especially *human* life ('anthropos' means 'human being' in Greek) – on earth as we know it is dependent on a number of fundamental cosmological constants which are closely attuned to one another.[71] It was discovered in the final quarter of the twentieth century that if a few of these constants (e.g., the strong coupling constant, the electromagnetic fine structure constant, the gravitational constant, etc.) had been just a fraction different from what they are, either smaller or larger, then intelligent life as we know it would have been completely impossible. The chance that the physical constants in question happen to be so remarkably fine-tuned as to support intelligent life on earth *just by coincidence* is unimaginably small. To express this chance graphically it has been compared with the chance that 'a tornado sweeping through a junk-yard might assemble a Boeing 747 from the materials therein'...[72] It is here that science moves most clearly in the direction of faith, for in such a situation it is much more obvious to accept a divine designer than to appeal to coincidence. Accordingly, Polkinghorne and many others assert that this strong anthropic principle makes them incline towards a theistic view of the world.[73]

We have provided only a small selection of the themes that can play in role in the contemporary dialogue between faith and (natural) science. In doing so, we have looked at faith in a rather general manner. Recent developments in science can also give us an insight into much more specific aspects of faith. So for example

70 See his 'Theism and Physical Cosmology', in: Philip L. Quinn & Charles Taliaferro (eds.), *A Companion to Philosophy of Religion*, Oxford 1997, 420. The term 'big bang' was first used by the astronomer Fred Hoyle, in a scathing remark during a debate on the BBC in 1949. Hoyle unambiguously rejected the Big Bang theory, in part because it did not fit well with his atheistic view of life.

71 Here the classic work is: John D. Barrow & Frank J. Tipler, *The Anthropic Cosmological Principle*, Oxford 1986, who refer to the constants in question as 'the *constants of Nature'* (5). For short summaries see Michael A. Corey, 'Anthropic Principle', in: Van Huyssteen (ed.), *Encyclopedia of Science and Religion*, 13-18, and McGrath, *Foundations of Dialogue*, 111-118.

72 Fred Hoyle in 'Hoyle on Evolution', *Nature* 294, no. 5837 (12 November 1981), 105.

73 John Polkinghorne, *The Way the World Is*, London 1983, 12-14, who also points out that the anthropic principle suggests a universe that must be as big as it is in order to make life on earth possible. Even Fred Hoyle is reported to have conceded in connection with the anthropic principle: 'Nothing has shaken my atheism as much as this discovery'; Denis Alexander, *Rebuilding the Matrix. Science and Faith in the 21st Century*, Oxford 2001, 414.

various writers on the Christology of Chalcedon, according to which Jesus was both truly divine and truly human, have thought afresh about this subject from the perspective of the so-called *principle of complementarity* of Niels Bohr. According to this principle, complex phenomena must sometimes be understood with reference to two apparently mutually exclusive models. Bohr's principle is especially concerned with light photons, which can be understood neither by exclusively using the so-called particle theory, nor by exclusively using the wave theory, but only by using both models. For the results of some experiments can only be accounted for by means of the particle model, whereas the results of other experiments can only be understood in terms of the wave model. Therefore, rather than opting for one of both models, we should keep them closely together. Of course Bohr's principle has to be adapted in various ways to make it applicable in the totally different context of Christology, but then it may nevertheless illuminate difficult conceptual questions regarding the person of Christ.[74]

Apart from the conversation between religion and the natural sciences, the dialogue model is also suitable in relation to other sciences. Thus, Barbour provides some examples of consonance between biblical data and recent theories in psychology, the neuro-sciences and anthropology.[75] Without doubt there will be further examples from other branches of scientific endeavour to add to this.

The examples we have discussed may in any case be enough to dispel an oft-repeated myth, namely that we should make no further attempts to relate God and religious belief to developments in the sciences, since this would only take us further and further back. The gaps in our knowledge would be increasingly filled in, and it would ultimately become impossible to find a place for God there.[76] The truth is, as we have seen above, that recent developments in the natural sciences do not make God redundant or the place for religious belief ever smaller (so that we no longer need the 'God hypothesis', as the French astronomer LaPlace once said to Napoleon), but rather it is precisely these recent developments which bring the question of God to the fore once again. Thus, we can agree with Del Ratzsch that '[h]istorically, science may have opened up as many gaps as it has closed'.[77] Here again it is far too simple to think in terms of a linear progress in science that goes hand in hand with a corresponding loss of ground for religious interpretations of the world. The picture is much more varied than this. It comprises

74 Cf. McGrath, *Foundations for Dialogue*, 187-206.

75 Barbour, 'Models and Relations', 763.

76 In this connection the pejorative phrase 'God of the gaps' is invariably used in the literature; its German equivalent has been coined by Dietrich Bonhoeffer: *Lückenbüsser-Gott*.

77 Del Ratzsch, *Science and Its Limits. The Natural Sciences in Christian Perspective*, Downers Grove, Ill. 2000, 181.

both conclusions which are drawn too quickly by believers as well as over-ambitious extrapolations made by supporters of the secularization thesis (which, briefly stated, holds that modernization always leads to a diminished significance of religion). Such simplifications must usually be retracted by both sides sooner or later.[78]

All in all, the dialogue model seems to be a particularly fruitful way of viewing the relationship between faith and science. In a certain sense, it is more open than the conflict model and its counterpart, the two-world model. It allows us to respect the specific nature of both faith and science. At the same time, it gives an impulse to creative multi-disciplinary research, the outcomes of which are not fully dictated by one's pre-existing convictions. *Perhaps* there is a conflict at some point, and *perhaps* the distinct identity of religion and/or science is overlooked at another issue – but there is no a priori reason for assuming that either of these necessarily has to be the case. Neither, therefore, is there any reason for stopping further inquiry into the concrete relationships between faith and an evolving science. Seen in this light, the dialogue model provides an enormous challenge to theology, which, after all, tries to think through the content and import of faith as systematically and comprehensively as possible. The question remains, however, as to whether the dialogue model is not too fragmentary to hold its ground in the long run. Should we not in the final analysis have as our goal a real *integration* of faith and science?

7.5 Attempts at integration

It is less easy than it perhaps seems to show the difference between the dialogue model and attempts at integration between faith and science. When we talk of a 'dialogue' this is in and of itself of course a metaphorical turn of phrase. In this way of thinking, faith and science are seen as two people or two groups of people who are in conversation with each other. The image presupposes the individual character and autonomy of each party. In the previous section, we indeed emphasized this distinctive character of both. However, a proper dialogue is never free-floating, but aims at concrete results to which both parties commit themselves. Accordingly, it may turn out that as a result of the dialogue between science and religion each of them has to some extent to accomodate itself to the other. When precisely this has to

78 Cf. e.g. Peter Berger's opening contribution in Peter L. Berger (ed.), *The De-Secularization of the World*, Grand Rapids 1999, 2: 'My point is that the assumption that we live in a secularized world is false. (…) This means that a whole body of literature by historians and social scientists loosely labeled „secularization theory" is essentially mistaken. In my early work I contributed to this literature.'

happen is difficult to say. How much ground did the Copernican system have to gain before the official Roman Catholic biblical exegesis of Joshua 10:12 had to be laid aside? In which year exactly did it become impossible to continue to hold to a Ptolemaic world view? Conversely, how strong must the anthropic principle be before we have to give up atheism? That can of course be heavily debated. But at some stage, it might simply become absurd to refuse this sort of shift any longer.

Now if, for example, we decide on the basis of the dialogue between science and religion that we should no longer believe that because of Joshua 10:12 we must subscribe to a geocentric world view, or if we decide, on the basis of the anthropic principle, that we should no longer believe that the universe came into existence by chance, then are we not as a consequence involved in the process of integrating faith and science? Barbour for one thinks we are. Hence, he discusses the anthropic principle in the section entitled 'integration'.[79] To be sure, Barbour distinguishes between, on the one hand, partial forms of integration (here, the anthropic principle is a case in point), and on the other a drive towards a complete *synthesis* between faith and science. However, Barbour himself implicitly admits that it is not completely clear where the dividing-line between the dialogue model and the integration model in fact lies. For example, he writes that 'advocates of Integration call for reformulations of traditional theological ideas that are more extensive and systematic than those envisaged by advocates of Dialogue,' thus suggesting that the difference between both models is a matter of degree.[80]

Indeed, no watertight distinction is possible and it is perhaps even the case that there is an inherent tendency to try to seek ever further integration. Anyone who, as a result of the dialogue between faith and science, has had to start revising their web of belief on a given point, will rather not want to make such a revision on too much of an *ad hoc* basis and will therefore to a greater or lesser degree attempt to seek integration. For in that way we minimize the chance that we are running after the facts. But then is it not best to seek a complete synthesis between them right from the beginning? Indeed, it seems that representatives of the dialogue model such as John Polkinghorne and also Ian Barbour himself are gradually moving in this direction.[81] The question we must ask though is what

79 See e.g. Barbour, 'Models and Relations', 764.

80 Barbour, 'Models and Relations', 763. It is interesting that Barbour talks here of the reformulation of *theological* ideas, thus suggesting that things only go in one direction here. To my mind, the example of the anthropic principle shows that things can also travel in the other direction, i.e. that current 'scientific' ideas can be re-formulated and traditional theological ones (e.g. that the universe is created) can rightly be affirmed.

81 With regard to Barbour, this is what Christian Berg argues in *Theologie im technologischen Zeitalter. Das Werk Ian Barbours als Beitrag zur Verhältnisbestimmung von Theologie zur Naturwissenschaft und Technik*, Stuttgart 2002, 70f.

price we have to pay for this. In any case there is an unmistakable difference in the character of the work of those who affirm dialogue, and in doing so continue to emphasize the distinct natures of faith and science, and that of those who affirm integration. The latter are often involved in developing some sort of all-encompassing metaphysical system.

As examples of *partial* integration, Barbour lists both natural theology and the theology of nature. These terms look similar but there is a clear difference between them. In a natural theology the starting point is the results of scientific research from which we, as it were, reason towards faith. The sources peculiar to faith, such as the special revelation of the Bible or the religious experience of people, are by definition not considered authoritative in the discourse of natural theology. In a theology of nature, by contrast, the starting point is precisely in the sources of faith, but then the idea is to align these carefully with the state of affairs in scientific research. In both cases we arrive at certain forms of integration, but in the case of natural theology the intention is always to affirm the classical tenets of faith (or part thereof) whilst in a theology of nature the intention is to adapt these to the changing circumstances.[82]

Natural theology has a long history. Its starting point is often associated with Thomas Aquinas, although on the one hand it goes back much further than this,[83] and on the other we must be careful in considering Thomas as a natural theologian. In any case he could not be called one in the sense that the biblical scriptures did not have any authority for him. He was though in the sense that he believed that the existence of God can be demonstrated by 'unaided reason', so apart from the Bible, but it is questionable whether he considered such rational demonstrations in a foundationalist fashion as the indubitable basis of faith (presumably, he believed in God anyway). Natural theology did become a fully-blown foundationalist project, however, in the 17th and 18th centuries under the impetus of the scientific revolution and the Enlightenment. In this context, it became a matter of primary importance to affirm the existence of God by means of reason and empirical observation, as these faculties were increasingly seen as forming the basis of all knowledge. Although many think that this project clearly failed as a result of the great critiques of religion in the nineteenth century (as above all put forward by Feuerbach, Freud, Darwin and Nietzsche), and others vigorously deny that we could in fact come to the true God in this way (Karl

82 It is interesting that Polkinghorne considers himself to be a representative of both and from this position attempts to integrate faith and science in an overarching perspective. For this, see Smedes, *Chaos, Complexity and God*, 24-27.

83 See Jaroslav Pelikan, *Christianity and Classical Culture. The Metamorphosis of Natural Theology in the Christian Encounter with Hellenism*, New Haven 1993.

Barth), there are still authors who continue the tradition of natural theology in sometimes impressive ways. One of the most well-known of these is Richard Swinburne, who in a series of books claims to have demonstrated on the basis of what we know from science that it is more likely that God does exist than that He does not. Moreover, Swinburne does not shirk the challenge to argue for more specific tenets of Christian faith in a similar way. Thus, in nother book he has argued that the chance that the gospel reports of the resurrection of Jesus Christ are true, if we take all the relevant facts into account, is exactly 97%.[84]

This is something that we will not find in the work of the Australian physicist Paul Davies. On the basis of his work on cosmology, however, he agrees with Swinburne's earlier conclusion that the idea of a cosmic Designer is more plausible as an ultimate explanation for the universe than naturalistic alternatives. He writes: 'Through my scientific work I have come to believe more and more strongly that the physical universe is put together with an ingenuity so astonishing that I cannot accept it merely as a brute fact.'[85] And Stephen Hawking had already given a hint in this direction when he finished his well-known book, *A Brief History of Time*, with the oft-cited conclusion that if we were able to provide a complete, coherent theory of the universe, we would then know 'the mind of God'. There has been some confusion about how to interpret this expression,[86] but in any case according to Hawking, natural scientists are already gradually finding such a complete, coherent theory of the universe. This complete theory would then give us a clear insight in God's ideas when creating the universe – or, alternatively, it might *substitute* what religious traditions describe as the creative mind of God.

If we then turn to what Barbour calls the *theology of nature*, things turn out to be somewhat different. Here the starting point, as we mentioned already, is not science but religion, and the aim here is to align religion as far as possible with discoveries in science in areas where these disciplines overlap. We see this most clearly in regard to the doctrine of creation. In addition, a huge amount of literature has been produced on the question of the extent to which the modern scientific

84 Richard Swinburne, *The Resurrection of God Incarnate*, Oxford 2003, 214; cf. Gijsbert van den Brink, 'How to Speak with Intellectual and Theological Decency on the Resurrection of Christ? A Comparison of Swinburne and Wright', *Scottish Journal of Theology* 61 (2008), 408-419. For a summary of Swinburne's earlier work, see his popular book *Is there a God?* Oxford 1996.

85 Paul Davies, *The Mind of God. Science and the Search for Ultimate Meaning*, London 1992, 16.

86 '... for then we would truly know the mind of God'. Stephen Hawking, *A Brief History of Time*, London 1988, final sentence. It seems that Carl Sagan, in his introduction to *A Brief History*, has given this a naturalistic slant and in so doing is responsible for the incorrect idea that Hawking advocates an atheist position in this book. See McGrath, *Foundations*, 40f.

picture of reality still leaves room for the activity of God *in the present day*.[87] Reflection on this often leads to a reformulation of the classical doctrine of providence. In both cases the revisions are above all characterized by the attempt to integrate the theory of evolution (or perhaps better: the evolutionary paradigm) into theology.

We come across an example of this in the work of someone we mentioned a little earlier, Arthur Peacocke (1924-2006), who, as a biochemist and theologian, concerned himself above all with the interface between theology and biology.[88] Whenever one reads Peacocke, one always gets a sense of the enormous complexity and stratification and subtle interaction of all sorts of biological processes in the world. Peacocke tries to fit his ideas about God as closely as possible into this broader picture. This means for him that we must not situate God and his work of creation away from or prior to the evolution of the universe, but rather right in the middle of it. God is present in the world as both creator and agent, in and through the whole process of evolutionary development towards ever more complex life forms. So God does not intervene in this process in the way that the classical doctrine of providence asserts, nor through chaotic systems or quantum events as Polkinghorne asserts. Rather, the whole evolutionary process is to be conceived of as a single glorious action of God. We can even say that this process more or less coincides with God. According to Peacocke, God created the world not outside himself, but rather in himself! For Peacocke all the stress lies on God's immanence, in other words on the fact that God is present everywhere in reality, even though He is not fully encapsulated in it. This means that God in fact manages to influence the world as it were from above (by means of what is called 'top-down causality'), but also that his omnipotence and omniscience are to a large extent limited by processes determined by chance and free will.

All this in fact requires a drastic revision of the traditional idea of faith, which not only affects the traditional doctrines of creation and providence, but also the classical image of God. According to Peacocke, God is even bound to the open process of evolution to such an extent that He cannot foresee its final

87 It is impossible to give even just a representative selection of this literature. I restrict myself, therefore, to giving a reference to the recent summary of N. T. Saunders, *Divine Action and Modern Science*, and to the multi-volume project of the Vatican Observatory in collaboration with the Center for Theology and the Natural Sciences in Berkeley; see on this Wesley J. Wildman, 'The Divine Action Project, 1998-2003', *Theology and Science* 2 (2004), 31-75.

88 For a discussion of Peacocke's work, see the recent issue of *Zygon* 43.1 (2008), which is dedicated to the significance of his work. For some interaction with Peacocke during the final stage of his life, see Philip Clayton (ed.), *All That Is. A Naturalistic Faith for the 21st Century*, Minneapolis 2007.

outcome. Anything could still happen in our reality. Thus, Christian eschatology, according to which the future is in God's hands, is also seriously qualified here, as well as Christology, according to which the world's future is secured at the deepest level by the work of Jesus Christ. The question we then have to ask is how partial the integration that Peacocke has in mind between faith and science really is. It seems that in his approach modern science influences and changes the classical content of faith to such an extent that it is in fact integrated seamlessly into the scientific world view. But then what we have is, in fact, a complete synthesis of faith and science.[89]

Meanwhile, the most influential explicit attempt to bring about such a complete synthesis is to be found in contemporary *process thought*, which arose from the philosophical system of the English-American mathematician and philosopher *Alfred North Whitehead.* It is of course no coincidence that Peacocke greatly admired process thought, and often refers to it in his work. For example, he shares with the process thinkers a panentheistic vision, according to which God and the world do not exactly coincide, but on the other hand cannot be thought of apart from one another, just as a person is not one and the same as his or her body, but also cannot be thought of separately from it. The most obvious difference between them is that process thought also tries to support the synthesis between faith and science philosophically by means of a completely new metaphysics, in other words a comprehensive, new 'picture' of how our reality looks at the most fundamental level.

Characteristic of process thought is the idea that the static categories that are determinative for traditional Western metaphysics, such as 'matter', 'essence' and 'mind', are replaced by dynamic categories such as 'process' and 'event'. At the roots of reality we do not find things or substances but rather events and experiences. Accordingly, it is not 'being' but 'becoming' that is the fundamental characteristic of reality. All existing entities are for Whitehead and his followers identical to 'actual events': they are continually in motion and subject to change. This change is on the one hand determined by other actual entities that exert influence, and on the other by the individual freedom of each actual entity itself. All entities have two poles, one passive (the physical aspect) and one active (the mental aspect). The more complex an entity is, the more developed its mental pole will be. But even the most simple entities have at least the germ of mentality, i.e.

89 On this section see above all Arthur R. Peacocke, *Theology for a Scientific Age*, London 1993. A more popular introduction to Peacocke's thinking can be found in his *Paths from Science towards God*, Oxford 2001. For a critical analysis of his ideas, see Smedes, *Chaos, Complexity, and God*, 75-119; Smedes especially objects to Peacocke's suggestion that theology must fully adapt itself to the sciences.

the possibility of self-determination. In this way the universe is developing second by second and at each point new possibilities are realised from what already exists.

This metaphysical world view is now also employed in the sphere of faith and theology, because according to Whitehead, God too is an actual entity. There is thus no qualitative ontological difference between God and creaturely reality. It is true that God encompasses all space and time. In contrast to other entities God is therefore eternal and omnipresent. But just like all other entities, God is 'dipolar'. God's concrete pole (or Consequent Nature, in Whitehead's terminology) is influenced by and responsive to other entities, and God's abstract pole (or Primordial Nature) consists in God's creativity and capacity for self-determination. As a source of creativity, however, God is only one agent amongst many. Therefore, God is not in a position to impose his will by force upon the world. His power is limited and has, by definition, to be realized by the exercise of influence and persuasion rather than by force. For clearly, every entity possesses a certain measure of freedom and creativity which cannot be compromised. The evident advantage of such a vision is that God does not apparently have to be held responsible for the many evils in the world (nor for everything that goes wrong in the great evolutionary process). On the contrary, God is in fact affected by evil, given that in his Consequent Nature God experiences all that happens. Accordingly, Whitehead was able to say that God is 'the great companion – the fellow-sufferer who understands'.[90]

It will be clear that such an over-arching vision has its great attractions. The consistency and subtlety with which it has been developed in every detail demand our admiration. Even in a postmodern period, which is suspicious of all forms of systems, it is possible to succumb to the appeal of this magnificent attempt to create one comprehensive synthesis in which faith is in complete accord contemporary sensibilities in culture and philosophy. Moreover, as we saw with Peacocke, process thought also offers us many possibilities to effect an integration between the different perspectives and goals of faith and science.[91] Incidentally, the same is also true for other comprehensive systems, such as that of the Jesuit and palaeontologist Pierre Teilhard de Chardin (1881-1955). Here too, metaphysical, theological and natural scientific lines of thought are woven together in a harmonious manner.

90 For a comprehensive development of his thought, see Alfred N. Whitehead, *Process and Reality. An Essay in Cosmology*, New York 1929; For this quote, see 532 (or 351 in the corrected edition edited by David Ray Griffin & Donald W. Sherburne, New York 1978). For the theological development that process thought has undergone, see John B. Cobb & David Ray Griffin, *Process Theology. An Introductory Exposition*, Philadelphia 1976.

91 In this regard, even Barbour has appropriated many of the ideas of process philosophy. See his *Religion in an Age of Science*, 218-242,

However, there are also great disadvantages and difficulties in this sort of attempt to achieve total integration. First of all, they have without exception a highly speculative quality, in part as a result of the new and totally specific jargon that has to be developed in order to reconcile such totally different fields as those of religion and science. Secondly, they often take us far from traditional faith convictions, in this case from classical Christian belief. The God of process thought seems to have little in common any more with the God that is described in the Old and New Testaments. In particular, this is a result of the fact that the transcendence of God is minimized and God's immanence maximized.[92] God no longer stands over the world, but in a very profound way is just as much a victim of the whole network of events in it as we often are. It is therefore unclear why this God would be worthy of our ultimate trust or even our worship and total surrender.

The most serious objection to *any* attempt at conceiving God as a continuation of our cosmos, and to any attempt to place both God and the natural world into one over-arching perspective is, however, that we would quite unavoidably be left with a pale and distant God who would neither criticize or judge our reality, nor act in a salvific or liberating way in it. In this regard a remark by Paul Davies is enlightening. He says that in the final analysis it is a question of taste and definition whether we call the deepest level of explanation of the universe 'God' or not.[93] In fact, if 'God' is just another name for the coping-stone of our attempts to explain everything, then it is difficult to see why we would give a hang about this God at all. It is nice to know how everything works at the deepest level, but we can then get back to our daily business. Our attitude to God would therefore be completely undemanding and totally indifferent!

Here it becomes clear that such a unifying way of thinking is at odds with the distinctive nature of faith. From the perspective of faith (which we will further elaborate in the next chapter), God is primarily the one who comes from the other side and whom we therefore cannot apprehend intellectually. God is the Transcendent One, or to use a more biblical term, the Holy One, who as the Almighty is also completely and utterly on the side of good and therefore worthy of our total devotion. Within process thought it is questionable whether God is really unequivocally good and holy. If it is true that God takes all our experiences up into himself and experiences them with us, then God will also be familiar with our experiences of moral evil.[94] Evil can then no longer be thought

92 That the process vision of God's transcendence is inadequate is also argued by Markus, *Beyond Finitude*, passim.
93 Davies, *The Mind of God*, 16.
94 See Markus, *Beyond Finitude*, section 8.4.

of as contrary to God's intentions. This is in fact an unavoidable consequence of seeing God as an extension of reality to the extent that process thought does.

In faith, however, we are not just dealing with reality as it is, but also with reality as it should be – or, to use the words of Willem Drees, not only with cosmology, but also with axiology.[95] From this point of view, God is not only the highest power, but also the highest good. If we ignore this, and all that we know is that God is the Creator or ultimate explanation of the universe, then we have not understood that faith is directed at a a life lived in the sight of God. For that reason the Christian doctrine of creation – as well as the doctrine of providence – should never be isolated from what is also believed about God in the Christian tradition: that God is the 'Law-giver' (and so the final standard of good and evil), the Saviour and the Re-creator of a reality which is not as it should be, and of people who do not live as they should live.

For this reason, when we are dealing with the relationship between faith and science the dialogue model should be preferred to attempts at integration, and we should resist the (utterly understandable) pull that takes us from the former to the latter. According to Barbour, this view is shared by the majority of writers in the field, because they too want to move beyond the conflict and independence models, but at the same time believe that we must content ourselves with dialogue between faith and science, possibly supplemented by partial forms of integration.[96] Apparently, many commentators sense the difficulties that are involved in seeking a comprehensive synthesis here. As I see it these difficulties mostly involve the constant threat of *hybris*, of over-confidence which loses sight of the human limitations in what are by definition vain attempts to fit a God who surpasses human thought into a comprehensive intellectual framework.

Finally, we also come to another conclusion. The lively debate which continues to increase in scope and hopefully also in quality about the relationship between faith and science shows that the idea of speaking responsibly about God in light of the current state of affairs in the (natural) sciences is definitely not out-dated. The findings in this chapter confirm the results of our reflections in the previous chapters of this study. For it became clear from the analysis of the developments in the philosophy of science that we carried out there that these developments do not exclude, but rather allow for, the possibility of a scientific theology in the classical sense of the word (i.e. 'speaking responsibly about God').[97] Our journey

95 Although I elaborate it differently from Drees, I readily borrow this twofold emphasis from him; see for example Drees, '„Religion and Science" without Symmetry, Plausibility, and Harmony', *Theology and Science* 1 (2003), 125-127.

96 Barbour, 'Models and Relations', 765.

97 See above, section 5.2 for this definition.

in the present chapter through various models of the relationship between faith and science confirms this conclusion. To be sure, a confirmation is not a verification and therefore never definitive. However, the burden of proof again lies with those who think that theologians and other believers, in response to the results of modern science, can at best give expression to their faith convictions in a quiet whisper.

8. Theology, church and society

8.1 More than science

Before summarizing the results of this study, in this closing chapter we first of all briefly return to the issue of the nature of theology. Much has already been said which should be said on this matter, but not yet all. For our attention has been focussed on the question of whether theology can be done in an academically responsible ('scientific') way. In the previous chapter, we found additional reason for answering this question in the affirmative, since it could be concluded there that the possibility of speaking responsibly about God has not been falsified by the current state of affairs in the (natural) sciences. Just like other sciences theology evolves from certain experiences which are undergone in the human encounter with reality. In the case of theology these are the experiences, both of a communal and a personal nature, which lead to the conviction that God has revealed himself in this world – according to Christians in a very tangible way in the history of Israel and Jesus Christ. Believers are deeply affected by this sort of personal and communal experiences, and interpret them in ways that give coherence to their views, expectations and life attitudes. They are, in one word, 'impressed by God'; and the (systematic) theologians amongst them in a certain sense derive their work from this status of being-impressed, by reflecting on it, seeking to get behind it, to clarify it and make it available for discussion. Although this happens in theology in a unique way given the specific nature of religious experiences, theological research is, as we have seen, structurally comparable with the way in which other scientists work out their 'being-under-the-impression' of given experiences or aspects of reality in their professional life.[1]

However, there is also a dimension in theology which breaks the analogy with the other sciences and which has remained underemphasized up to now in this book, namely what we might call a spiritual dimension. It is often suggested that as a result of this spiritual dimension theology as a whole falls short of other the scientific academic disciplines. More positively, its spiritual dimension is

1 For an account of religion as 'being overwhelmed by God' in the midst of the multiple overwhelmings of (modern) life, and of academic theology as being located at a convergence of these overwhelmings, see David F. Ford, *Theology. A Very Short Introduction*, Oxford 1999, 7-12.

sometimes considered as proof of the fact that theology breaks through the well-defined framework of the other sciences and is superior to these. It is not necessary, however, to pass judgement on the impact of this spiritual dimension of theology in either of these ways. It is much more important to identify it as clearly as possibly and not to ignore it.

In the history of theology itself this extra-scientific aspect of theology has often been denoted by the term *sapientia* (wisdom), a term which was used consciously in contrast to *scientia*. We find this tradition, for example, in Augustine, according to whom knowledge (*scientia*) is always related to earthly and temporal things, whilst contemplation (*contemplatio*) of eternal things on the other hand, in particular of God as the highest good, leads to *sapientia*.[2] Augustine did not intend here to play one off against the other or to undervalue scientific knowledge; the sciences can serve wisdom just as well and indeed lead to it.[3] But as soon as we start dealing with God, a dimension comes into play which goes beyond scientific knowledge. Augustine identifies this dimension in terms of piety, love and worship. The knowledge of God can, in short, never be separated from a particular *attitude*. And here it differs structurally from scientific knowledge of all sorts of every-day objects.

In the twelfth and thirteenth centuries with the establishment of the first universities we find the process taking place whereby theology came to be seen as one of the sciences. But even then many people (for example, Bonaventura, Duns Scotus, Luther) continued to call it at best a *practical* science and not, as Thomas Aquinas demanded, a speculative one.[4] If we look at how this notion of a practical science was developed, then it turns out that something not very different was meant from what was covered by the term *sapientia*. In both cases the central idea was that it is impossible for human knowledge of God to be 'existentially neutral'. Rather, our knowledge of God forms an inseparable unity with the adoption of the appropriate attitude of faith, love and honour towards God. Following on from this, Luther could even say that the true object of theology is not God, but *the human being*. Theology is about how we sinful human beings are

2 Augustine, *De Trinitate* XII, 14, 21. Before Augustine, Clement of Alexandria had already characterized Christian doctrine as wisdom (i.e., as that which all philosophers – lovers of wisdom – are, without knowing it, engaged in seeking). In the background 1 Corinthians 1:30 played a role here. For later elaborations of this theme, see e.g. M. Hailer, *Theologie als Weisheit*, Neukirchen 1997.

3 For this, see Wolfhart Pannenberg, *Theology and Philosophy of Science*, 9-12, and the further references given there.

4 Pannenberg, *Theology and Philosophy of Science*, 231-235.

justified by God.[5] By means of this 'relational' definition of theology, Luther gave a new twist to the practical character of theology.

Accordingly, in theology we are definitely not *only* dealing with 'responsible talk about God'. On closer inspection, this definition of theology (as it was used a couple of times in the previous chapters) turns out to be incomplete. For when we talk about God we simply cannot do so in an isolated way. In theology it is as much about human beings, creation, the church, society etc., and in each case from the point of view of their *relationship* to God.[6] Theology never deals with God-as-such, but always with God who speaks to us, and who relates to us and our universe in many other ways. The proper object of theology, therefore, is this relationship between God and everything else. Thus, God is not an object like other objects and so cannot be approached in the same way as the common objects of scientific research. Vincent Brümmer clearly articulates the difference as follows:

> Truth claims cannot be eliminated from religion. Yet, such claims differ in fact from those in science by being logically related to the religious way of life. (…) For this reason they are 'existential' in a way that the truth claims of science are not. It might make sense to say: 'It is true that the planet Jupiter exists and is the largest planet of our solar system, but I don't really care much about that'. It is however absurd to say: 'It is true that God exists and is the personal creator of the universe, but I don't really care much about that' Here (...) Wittgenstein might comment: 'For a blunder, that's too big'.[7]

Thus, theological statements always imply a certain way of life. In this sense theology is indeed a 'practical science', with the practical goal of living rightly in God's sight. Theologians of an anti-speculative persuasion such as Peter Rames, William Ames and Johannes Cocceius preferred to define theology in such practical terms. They literally described theology as 'a doctrine for living well', 'a doctrine of living for God' and 'a doctrine which is in accordance with piety'.[8]

In doing so, it was not their intention to deny any orientation towards knowledge, but rather to highlight the special character of the knowledge that we find in theology. The Heidelberg Catechism has articulated that special character in a way that is unequalled, by describing faith as 'not only a certain knowledge (…)

5 'Theologiae proprium subiectum est homo peccati reus ac perditus, et Deus iustificans ac salvator hominis peccatoris.' Luther *WA* 40, II, 328, 17 (with reference to Psalm 51:2). See Pannenberg, *Theology & Philosophy of Science*, 235.

6 So Thomas Aquinas, *Summa theologiae* I 1, 7. Cf. Herman Bavinck, *Reformed Dogmatics*, Volume 1: Prolegomena, Grand Rapids 2003, 36.

7 Vincent Brümmer, *The Model of Love*, 18 (the Wittgenstein-quotation is from *Lectures and Conversations on Aesthetics, Psychology and Religious Belief*, Oxford 1966, 62).

8 Resp. 'doctrina bene vivendi', 'doctrina Deo vivendi', and 'doctrina, quae est secundum pietatem'; for the references, see Willem J. van Asselt, *The Federal Theology of Johannes Coccejus (1603-1669)*, Leiden 2001, 28, 75.

but also a wholehearted trust' (Sunday 7).[9] This knowledge of faith which is full of confidence is nourished by an intimate intercourse with God through prayer and reading Scripture, and above all through the appropriation of the promises contained in it, which results in an attitude of gratitude.

We can therefore concur with Alan Padgett, who in his study on theology and science characterizes the *praise of God* as the final aim of theology. Padgett shows that theology has in many cases become isolated from the human longing to know God in a personal way.[10] Taking his lead from others, Padgett ascribes this to the intellectualization of theology as a result of its being situated within the academy, and he tries to overcome this dichotomy by orientating himself anew towards the original goal of theology.[11] He argues that it is present-day postmodernity with its rejection of the intellectualism of the Enlightenment which offers new possibilities for this. Indeed, we could see the appeal that is heard nowadays in many theological colleges to give more attention to spirituality in their programmes of study as an expression of this changing climate. This appeal will however have to be heeded across the board and not only in the introduction of courses in spiritual formation or by bringing in special lecturers who concern themselves with spirituality. Otherwise it would remain more of an expression of the seriousness of the problem rather than the start of a process to address it.

The concept of theology that we have been articulating in the previous chapters gives theology the opportunity to be related once more to its original spiritual dimension. H. J. Adriaanse had to admit at the end of his reflections on the nature of a scientific theology, 'Scientific theology is silent about God. Personal belief stays out of play.'[12] In the meantime, however, as we have said, the conditions in the philosophy of science have changed. We can think for example of the increased awareness of the role that perspectivity and (inter)subjectivity – whether or not articulated in terms of scientific paradigms – play in every form of scientific endeavour (chapter 6), and of the great potential of the dialogue model for developing the relationship between religion and science (chapter 7). Such developments point

9 Cf. Jaroslav Pelikan, *Credo. Historical and Theological Guide to Creeds and Confessions of Faith in the Christian Tradition*, New Haven 2003, 45f.

10 Padgett, *Science and the Study of God*, 105 (Padgett speaks of the 'spiritual and religious quest to know God in a personal way').

11 *Science and the Study of God*, 106. See also Geoffrey Wainwright, *Doxology*, New York 1980, and Dan Hardy & David Ford, *Praising and Knowing God*, Philadelphia 1985. For an analysis of the background to the separation of piety and science in theology, see Edward Farley, *Theologia. The Fragmentation and Unity of Theological Education*, Philadelphia 1983.

12 Adriaanse et al., *Het verschijnsel theologie. Over de wetenschappelijke status van de theologie*, 132.

into one direction: the possibility of a scientifically responsible way of speaking about God can no longer be ruled out a priorily. As a result, we no longer have to agree with the view that a scientific theology has to be silent about God. Having re-established the connection between academic theology and (speaking about) God, however, the way is open once more for a new relationship between theology and spirituality. The old ideal of the Dutch scholastic theologian Gisbertus Voetius, who long ago looked for the right way to relate science and piety to one another,[13] then comes back to the fore.

But Voetius was not the only one. Many great thinkers in the history of theology have pointed implicitly or explicitly to the huge importance of a spiritual life consisting of prayer and meditation for theology. Augustine and Anselm even wrote some of their theological treatises in the form of a direct personal address to God – and not by way of pious flourish, but intentionally so. Thomas Aquinas opened his *Summa Theologiae* with a prayer for the right attitude in his study, so that it would be to the honour of God's name. Luther was of the opinion that theology could only be carried out from real-life experience (*experientia*); for him, 'to know' (*cognoscere*) means 'to feel through experience' (*sentire experientia*)[14], and accordingly in a famous formula Luther could summarize the correct way of doing theology in three rules: prayer, meditation and spiritual struggle (*oratio, meditatio, afflictio*).[15] In a similar vein John Calvin asserted that all right knowledge of God is born out of obedience.[16] In the twentieth century, Bonhoeffer considered prayer to be the beginning of theology[17] and Barth as well insisted that theology can only be performed in the act of prayer.[18]

However, all this is not to deny that theology in a certain sense also has to stand at a distance from the spiritual life. For clearly, theology is a second order activity: it reflects upon the first order language as it is used in the practice of

13 G. Voetius, *De pietate cum scientia conjungenda* (inaugural address, Utrecht 1634).

14 Luther, *WA* 40/II, 328.

15 *Preface to the Wittenberg Edition of Luther's German Writings 1539;* for an English translation, see e.g. Luther's Works, American Edition, Vol. 34. Philadelphia 1960, 285-288. Cf. for an extensive explanation Oswald Bayer, *Theology. The Lutheran Way*, Grand Rapids 2007, 33-66 (where Luther's practical understanding of theology seems to be played off too much against that of theologians working in the tradition of faith-seeking-understanding, 34).

16 '… no one can get even the slightest taste of right and sound doctrine unless he be a pupil of Scripture. (…) all right knowledge of God is born of obedience.' John Calvin, *Institutes* I 6, 2 (tr. F.L. Battles, 72); the last sentence is quoted with approval by Karl Barth, *The Doctrine of the Word of God*, Edinburgh 1975, 19.

17 In his Berlin lectures; see *Berlin 1932-1933* (Dietrich Bonhoeffer Werke, Volume 12), Gütersloh 1997, 160 (an English translation of this volume is forthcoming).

18 Karl Barth, *Evangelical Theology: An Introduction*, Grand Rapids 1963, 160.

faith, for example in worship and prayer, and it tries to clarify it, to analyze it critically and to account for it. Therefore, we cannot agree with Padgett when he describes theology without qualification as a form of liturgy and worship.[19] Then non-theologians would be missing a lot! The suggestion that theologians by virtue of their profession are closer to God than others has in the past often led to protests, and rightly so. In reaction to this, others tend to separate theology from any spiritual orientation towards God, giving it a much more modest task (the study of aspects of faith and religion for example)[20] – but the consequence of this is that scientifically responsible talk of God is at risk once more! To my mind there is no justification for such a drastic reaction, but theologians will do well to avoid any appearance of self-aggrandizing in this regard.

Padgett tries to do this by labelling all sorts of first order activities as theology too.[21] As a result, every believer is considered to be a theologian (and not just an amateur theologian, but a theologian in the true sense of the word). However, this implies a somewhat idiosyncratic and undesirable attenuation of the meaning of the term 'theology'. It is better to assert that theology is *nourished* by faith, springs from faith, remains in (critical) harmony with it and above all stands in the service of faith; but still it is quite a different activity from the practice of faith. Theological language is of another order than the language of faith, because it reflects on the language of faith and inquires into its meaning and truth. Along with others we could say that theology is a form of 'serving God on the field of thought',[22]or of 'loving God with our minds'.[23] In such expressions we come across the coincidence of spiritual warmth and nearness ('serving', 'loving') and a certain distance ('field of thought', 'our minds'). Meanwhile, we must bear in mind that theology is not the only way in which we can fulfil this commandment to love God 'with all our mind'. For example, being involved in one of the (other) sciences can be an equally valid way of loving God with our mind. Even so theology

19 'Theology (…) is an integral aspect of the liturgy, and of the worship and praise of God'; which is summarized a little later as 'theology is worship'. Padgett, *Science and the Study of God*, 108.

20 See for example W. J. Ouweneel, *Christian Doctrine. Old Truths in New Perspectives* 1, Amsterdam 1995, 51, 71. It is interesting that the supporters of this view, who are usually orthodox Protestants, are close to liberal thinkers on this point. On the basis of agnostic considerations, the latter however understand theology merely as the study of religion.

21 'Prayer, sermons, hymns, worship, and liturgy, are already theology'. Padgett, *Science and the Study of God*, 108.

22 Thus Karl Barth; cf. Eberhard Busch, *The Great Passion. An Introduction to Karl Barth's Theology*, Grand Rapids 2004, 13.

23 Thus Hendrikus Berkhof, *Introduction to the Study of Dogmatics*, Grand Rapids 1985, 14.

is a very special, pregnant way of reflecting on God's loving and gracious activity that kindles our faith and precedes all our talk of God.

Theology in this sense is therefore more than science. For even though we can point to this sort of faith moment in all scientific activity, and even though those who practice science are almost by definition in a certain sense taken over by and filled with enthusiasm about the object and field of their discipline, that is still something different from the faith commitment of theology that we are talking about here. Given the current state of affairs in the philosophy of science it would be artificial and hardly convincing to stretch the concept of science so far that the spiritual dimension of theology could be included in it. So theology necessarily has an extra-scientific spiritual dimension. This dimension, however, does not stand in the way of its scientific character. It is no excuse for coming up short intellectually or for laziness of thought. On the contrary, it should stimulate theologians to do their work as responsibly, critically and passionate as they can.[24]

All in all, we maintain that theology must be situated between faith and science. Theology is the bridge between the two of them, and for this reason has piers on both banks of the river. On the one hand when it comes to its structure, it has much in common with the other sciences. On the other hand it has much in common with the faith from which it springs and by which it (or rather, the theologian!) is continually nourished. In this way, theology participates in the existential, relational and spiritual character of faith, even though its tone (just as with all science) can be abstract and at times even replete with technical language.

Because we have concentrated on the questions concerning theology-as-science, the close relationship between theology and faith has been underexposed in this study. For this reason we briefly wanted to give it some attention – not as an addendum that must necessarily remain an erratic block in the concept of theology that was developed, but as an enriching perspective that is opened up precisely from within this concept.

8.2 Theology as a public concern

Finally, we shall try to draw together some threads which run through this study by coming back to the different public functions that theology has. One of those who has placed attention for the public character of theology firmly on the agenda is the influential American theologian David Tracy (born 1939). Tracy has rightly pointed out that theologians who take the radical theocentrism of the

24 Cf. here, once again, Karl Barth's notion of theology, which is a very critical notion – but precisely in the service of faith; cf. his *Doctrine of the Word of God*, §1.

Judaeo-Christian tradition seriously can in no way be comfortable with the perilous modern view that religious convictions are only personal preferences or private choices. If God is indeed the Creator and Sustainer of the world, then the reality of God is universal in nature and speaking of God also has a universal tenor. Speaking-of-God therefore always means speaking *publicly*, and intentionally so.[25]

In this connection Tracy differentiates between three public domains or forums to which theology ought to direct its work: that of the church or faith community within which we operate, that of the academy and that of society in the broader sense of the word.[26] Tracy observes that theologians often limit themselves to one of these three domains. For example, they may concern themselves predominantly with society without paying much attention to the church – or vice versa. They expend all their effort in trying to satisfy academic standards, or they tend to ignore such standards because of their wish to be loyal to a particular church or to society. However, Tracy rightly argues that for a public theology it is a important to work with *all three* of these forums in order to be able to speak to every sort of audience in an appropriate and inspiring way. Tracy develops this in such a way that in so-called fundamental theology (or, to use the Protestant term, in the prolegomena of dogmatics) a certain correspondence with the academic world is sought, whilst systematic theology deals with the church and practical theology is focussed in particular on society and its problems. It can be questioned, however, whether theology's responsibilities can be or should be divided so neatly among its constitutive parts. Rather, it is important to grasp that in fact there is no contrast between the three responsibilities. For example, theologians usually best serve their church community by striving after the highest academic standards in their work, both in terms of research and in terms of teaching. As David Ford says, '[c]hurches and other faith communities need high-quality studies and discussions of issues relevant to them, and they also need members (...) who are well educated and theologically literate'.[27] Likewise, theologians serve the academy by showing the relevance of their work for society at large. And they serve society by bringing in religious and theological concerns in debates about public issues concerning economics, politics, the media, education, family life, law, medicine, et cetera.[28]

25 David Tracy, 'Defending the Public Character of Theology', *The Christian Century* 98 (1981), 350-356.

26 David Tracy, *The Analogical Imagination*, New York 1981, 21. Tracy talks of three 'publics'. Similarly, David Ford, *Theology*, 19, discusses the worldwide academic community, the community of faith and society as a whole as theology's 'three basic orientations of responsibility'.

27 Ford, *Theology*, 19.

28 Ford, *Theology*, 19.

Now according to Tracy the criteria for what can be said meaningfully in fundamental, systematic and practical theology ought to be derived from the second and above all the third of the three public domains that he distinguishes: 'The theologian should argue (...) on strictly public grounds that are open to all rational persons'.[29] Following Alan Padgett, we believe that on this point Tracy is in part right, but in part wrong. Where he is right is that theological work ought indeed to be understandable and clear so that people within the second and third publics can follow what is going on. However Tracy is too naïve if he thinks that there is a sort of universal human rationality which should count as the ultimate source of truth and meaning and which should therefore form the only authority to which theologians should appeal.[30] For such an authority is unmasked as a myth by postmodernity, and theologians can therefore (just like anyone else) do nothing other than take as a starting point the particularity of their own tradition.[31] In the case of the Christian theologian this means that he or she ought to orientate themselves towards the sources of their own tradition, that is to the biblical witness of what they see as God's revelation – in order to articulate its meaning and significance in as accessible a manner as possible. We saw in section 4.3 that in this regard there is no a priori opposition between a faith perspective (appealing to revelatory experiences) and a 'rational' approach, given the fact that in the latter as well some form of perspectivity is always involved.

However, even though Tracy, in working out his plea for a public theology, remains too caught up in the framework of the Enlightenment, we should take this plea as such very seriously. And its focus on the three forums of the faith tradition, the university and society seems to me to be still relevant, especially in a time in which we experience an unfortunate splitting up of church-bound theological institutions on the one hand and academic departments of religious studies on the other. David Ford rightly argues that examples of 'best practice' in the field are to be found at those faculties, departments or institutions which try to overcome this bifurcation by combining descriptive religious studies and constructive theology with each other in ways that are mutually enriching.[32]

29 *The Analogical Imagination*, 64. For Tracy this means that an appeal to (special) revelation, as is usual in Christian theology, is not justified.

30 Padgett, *Science and the Study of God*, 112. For a similar criticism, see D. A. Carson, *The Gagging of God. Christianity Confronts Pluralism*, Leicester 1996, 80f.

31 Incidentally, Tracy has developed an ambivalent relationship towards postmodernity, which it is not so easy to give an account of. See for example his 'Theology and the Many Faces of Postmodernity', *Theology Today* 51 (1994), 104-114, and confer Carson, *Gagging of God*, 80, 100f.

32 Ford, *Theology*, 17.

It is true that Padgett tries to outdo Tracy's proposal by stipulating that the first and foremost 'public' to which theology ought to attend is neither the academy, nor the church nor society but God.[33] This is a suggestion which surely has some initial plausibility. Theologians should indeed constantly be aware that they perform their activities *coram Deo*. This will encourage them do their work with integrity and a proper consciousness of their responsibility. However, as an alternative or addition to Tracy's proposal, Padgett's suggestion simply will not do. For clearly, consciously being active 'in the sight of God' ought not to be something apart from the theological concern for church, science and society. On the contrary, it should rather intensify this concern. God himself does not need our theology, and in this sense does not form a separate 'public' along with the others. But the fundamental loyalty of the theologian towards God does stimulate his of her efforts towards a public theology. God is of course the God of the whole world – or He is no God at all. Faith in God therefore implies hope for the church and for the world as created reality, and so concern for both. Tracy himself emphasized this relationship between the universal character of Christian faith in God and the tasks of theology, and in this way justified his theory about the three 'publics of theology' theologically.[34]

But this does not take away the fact that with regard to the typology developed by Tracy one can also set about things differently. Thus, for example, Rowan Williams put forward an alternative triad, namely of 'celebratory', 'communicative' and 'critical theology'.[35] The first arises from the canonical and confessional writings together with the liturgy and wealth of hymns of the faith community to which the theologian belongs. This proclamatory form of theological practice stands in the tradition which connects theology with wisdom and with a practical way of life, a tradition that as we have seen has a long track record in history. It is also closely related to the personal living out of faith and is suffused with a warm spirituality. Communicative forms of theology by contrast use another, more general, language register, because they address themselves to the conversation with the cultural and intellectual environment of the church. Finally, critical theology evolves whenever that conversation becomes so intense that the foundations of the faith tradition are at issue. In a critical theology, we then either have to revise this tradition fundamentally or raise the drawbridge to gen-

33 'The Blessed Trinity is the first public for our theological literature'. Padgett, *Science and the Study of God*, 111.

34 'An insight into the universal character of the divine reality as the always-present object of the Christian's trust and loyalty is what ultimately impels every theology to attempt publicness'; Tracy, 'Defending the Public Character of Theology', 354.

35 Rowan Williams, *On Christian Theology*, London 1999, xiii-xvi.

eral culture.[36] The advantage of Williams' typology is that the transition from one type to another can be thought of fluidly here. Williams in fact draws a continuous line from the heart of Christian theology (which is situated in worship) to the furthest periphery of it, and all sorts of clearly developed theological positions can be marked in somewhere along the line. Even the most remote critical positions can be (charitably) interpreted as being at the service of the faith community.

In what follows, however, we shall take our lead from the more clear-cut markers used by Tracy and Ford when they distinguish between the forums of church, academy and society. For what we have tried to do so far is in fact to relate theology to these three forums.

To begin with we turned our attention to *society*, and we tried to carry on doing this for a few chapters without letting this perspective be interfered to any great extent with considerations which are internal to theology. We discussed society in particular from one point of view, namely the enormous changes that Western culture, but also in increasing measure global culture, has undergone and is still undergoing as a consequence of the wide-reaching effects of science (chapter 1). This brought us to a point where we subjected this phenomenon of science to closer analysis. In particular we asked about the extent to which the common-or-garden picture that most of us have of science is supported by the findings of those who study this phenomenon professionally: philosophers of science, historians of science, and other practitioners of 'science studies'. And this turned out not to be the case in at least three crucial respects.

First of all, the development of science turns out to go in a much less linear-progressive way than is usually supposed to be the case from the standard picture of science, which is largely based on the idea of progress. Even though we did not dispute the idea that in the development of technology we can talk of a more or less constant progress, it turned out that this often causes us to lose sight of the many accidental twists and turns, fault lines, deviations and dead-ends that characterize the development of the natural sciences. There is certainly no guarantee that we are on the right track, or that we are getting closer and closer to our destination – a comprehensive knowledge of the universe. We saw how, in terms taken from Kuhn and Lakatos, at crucial moments, our observations turn out to be theory-laden, whereas our paradigms and research programmes take it in turns

36 An example of the first strategy is the so-called Chicago School in the USA, of which amongst others David Tracy is a representative. An example which comes near to the second is the Yale School, with theologians such as Hans Frei and George Lindbeck. See David Ford (ed.), *The Modern Theologians*, Oxford 1997, 343-356 (a section which was not reprinted in the third edition, Oxford 2005).

with one another on pragmatic grounds. As a result, we are unable to determine objectively if and how we get closer to the truth (chapter 2).

Secondly, to talk of 'science' turned out to be in a certain sense somewhat deceptive. The phenomenon of science is in fact quite pluriform and multi-coloured, so that it is more correct to talk about it in the plural. We have tried to become more sensitive to the multiplicity of sciences and the large variety of their methods, without a priorily ascribing a superior position to the natural sciences. Furthermore, we saw that in one respect the sciences are no different from the humanities, i.e. in their fundamental aim, which is to understand and interpret reality as adequately as possible. At the same time we discovered that this goal must be pursued in very different ways according to the nature of the specific research field of a given science. We therefore concluded that science is more of a 'family concept' than one which we can define in a clear, unambiguous way (chapter 3).

Thirdly, the current idea that in science things go 'from fact to theory' turned out to be incorrect. We examined what is understood by the theory-laden nature of perception, and discovered how the principle of perspectivity that plays a decisive role here affects the practice of science on all sorts of levels. The theory (which is still very influential) according to which it is possible to draw a neat dividing-line between 'objective' science and 'subjective' prejudices turned out to be untenable. Even the 'hard' notion of scientific *proof* turned out on closer inspection not to be an exception to the intersubjective nature of the scientific endeavour. Science is practiced by people who are influenced in all sorts of subtle and less subtle ways by the particular society of which they form a part, and the contingent perspective from which they think and act can never be totally filtered out of the results of their work. Neo-marxist critics, feminist theoreticians and a distinctive philosopher of science such as Feyerabend elucidated the extent to which this is the case (chapter 4).

After having examined in this way a particularly determinative dimension of modern society, we then turned our attention to Tracy's second forum, that of the *academy*. Here we brought theology more emphatically into play. For a theology that wants to be relevant in the public domain, a recognized place in the academy is of great importance. The question which is decisive here is whether theology may be called a science (by which we meant in this connection: a proper academic discipline). First of all we dealt with this question from the perspective of what we referred to as the 'older philosophy of science', and in doing so we came to some sobering conclusions. Taking logical positivism and critical rationalism as our intellectual frameworks, the possibilities for 'responsible talk of God' are naught or next to naught. In passing, we saw that the older frameworks of the philosophy of science still influence the actual practice of theology and its results to some extent. This leads to a real diminution of the meaningfulness of classical theology and to an atrophy of vital Christian intuitions (chapter 5).

Subsequently we examined in detail how theology stands given the state of affairs in the newer philosophy of science. Things turned out to be much more favourable here. In Pannenberg's work, we found that at a very early stage of the newer developments he moved back and forth between the older and the newer philosophy of science. As a result, the idea of theology as a science became once more possible with him, but it was bound by such fundamental restrictions that the question arised as to whether in order to operate as an academic discipline theology did not have to deny its very nature and object. Almost two decades later, however, we heared a more assertive voice in the work of Murphy. Now that the newer philosophy of science had become firmly established, theologians' self-confidence had increased. But didn't Murphy think all too quickly that she had cracked the problem? At least at one crucial issue it seemed that she underestimated the seriousness of theology's predicament. This led us to an attempt to sketch the contours of another blueprint for a scientific theology (chapter 6).

After this it was time to turn our attention to Tracy's third public: that of the *church*, or, in a broader sense, the Christian faith tradition. How do they relate to the present-day sciences and their results? The theory, that they are by definition in conflict with science – a conflict that the church of course must lose ever more clearly – could not be confirmed from the available material. On the contrary, we found that this 'conflict model' is decidedly left behind at the moment by the majority of religion-and-science scholars. Even the so-called independence model, according to which faith and science ought to be kept carefully apart from one another because they give answers to categorically different questions, turned out to be inadequate in certain respects. This was not the case for the dialogue model, which in the current discussion, and not without reason, has far and away the greatest number of advocates. Attempts to integrate faith and science into one huge overarching concept, however, tend to evoke many hard questions. In all likelihood such a uniform all-comprehensive view will remain unattainable given the limited character of our human capacities for understanding (chapter 7).

Finally we examined the relation between theology and the Christian faith tradition more closely by briefly focusing on the spiritual dimension of theology. It was argued that the rootedness of theology in a communal and personal faith that is lived out need not detract from its scientific nature. This important spiritual dimension of theology, which has to do with personal faith and trust, with prayer and worship, has a pre-scientific character which may fuel academic theology rather than being at odds with it. We did not though develop the idea of how such a warm, 'experiential' theology, which is at the same time appropriate from a scientific perspective and public in nature would look like. In the framework of this book we could only give a brief indication of it (chapter 8).

All in all, we conclude that the recent developments in the philosophy of science and the philosophy of culture resulting from the contemporary demise of positivism and Enlightenment thought give no reason for unbridled triumphalism. The current contentiousness (or *Strittigkeit* as Pannenberg calls it) of Christian faith in today's society is too serious for that. However, these developments do justify a new passion for advocating, with proper confidence, the Christian message as a truth which can be experienced in the public domain of today's culture.[37] Whenever theology steps back from this task, its sphere of influence will quickly be limited to an ever smaller private realm, a process that will leave the church as one religious sect among many others. If anything, this book has shown that there is no need for such a withdrawal.

37 Compare the titles of two of Lesslie Newbigin's latest books: *Truth to Tell. The Gospel as Public Truth*, London 1991; and *Proper Confidence. Faith, Doubt and Certainty in Christian Discipleship*, Grand Rapids 1995.

Index

Bibliography

Adorno, Theodor W., et al., *The Positivist Dispute in German Sociology*, London 1976 (or. German edition: Berlin 1969).

Adriaanse, H.J., *Vom Christentum Aus. Aufsätze und Vorträge zur Religionsphilosophie*, Kampen 1995.

Adriaanse, H.J., H. A. Krop & L. Leertouwer, *Het verschijnsel theologie. Over de wetenschappelijke status van de theologie*, Amsterdam 1987.

Albert, Hans, *Treatise on Critical Reason*, Princeton 1985 (or. German edition: Tübingen 1968).

Alexander, Denis, *Rebuilding the Matrix. Science and Faith in the 21st Century*, Oxford 2001.

Alston, William P., 'Christian Experience and Christian Belief', in: Alvin Plantinga & Nicholas Wolterstorff (eds.), *Faith and Rationality*, 103-134.

– *Perceiving God. The Epistemology of Religious Experience*, Ithaca 1991.

Aquinas, Thomas, *Summa theologiae*.

Asselt, Willem J. van, *The Federal Theology of Johannes Coccejus (1603-1669)*, Leiden 2001.

Audi, Robert (ed.), *The Cambridge Dictionary of Philosophy*, Cambridge 1995.

Augustine, *De Trinitate*.

Austin, J.L., *How to Do Things with Words*, Oxford 1962.

Ayer, Alfred J., *Language, Truth and Logic*, Harmondsworth 1971 (orig. ed.: New York 1935).

Banner, Michael, *The Justification of Science and the Rationality of Religious Belief*, Oxford 1990.

Banner, Micheal & Alan Torrance (eds.), *The Doctrine of God and Theological Ethics*, London 2006.

Barbour, Ian G., *Myths, Models, Paradigms*, New York 1974.

– *Religion in an Age of Science*, San Francisco 1990.

– *Religion and Science. Historical and Contemporary Issues*, San Francisco 1997.

– *When Science Meets Religion*, London 2000.

– 'Science and Religion: Models and Relations' in: Van Huyssteen (ed.), *Encyclopaedia of Science and Religion*, 760-766.

Barrett, Peter, *Science and Theology since Copernicus. The Search for Understanding*, London 2004.

Barrow, John D. & Frank J. Tipler, *The Anthropic Cosmological Principle*, Oxford 1986.

Barth, Karl, *The Doctrine of the Word of God* (Church Dogmatics I,1), 2nd edition, Edinburgh 1975 (or. German edition: Zollikon-Zürich 1932).

– *The Doctrine of Creation* (Church Dogmatics III.1), Edinburgh 1958 (or. German edition: Zollikon-Zürich 1945).

– *Evangelical Theology: An Introduction*, Grand Rapids 1963 (or. German edition: Zürich 1962).

Bartley, William W., *The Retreat to Commitment*, London 1964.

Bavinck, Herman, *Reformed Dogmatics*, Volume 1: Prolegomena, Grand Rapids 2003 (or. Dutch edition: Kampen 1928).

Bayer, Oswald, *Theology. The Lutheran Way*, Grand Rapids 2007.

Beek, A. van de, 'Is God Detectable?', in: Henri A. Krop et al. (eds.), *Post-Theism. Rephraming the Judeo-Christian Tradition*, Leuven 2000, 209-221.

– *Ontmaskering. Christelijk geloof en cultuur*, Zoetermeer 2001.

– *Jesus Kyrios. Christology as Heart of Theology*, Zoetermeer 2002 (or. Dutch edition: Kampen 1998).

– *De kring om de Messias. Israël als volk van de lijdende Heer*, Zoetermeer 2002.

Belt, Henk van den, *The Authority of Scripture in Reformed Theology. Truth and Trust*, Leiden 2008.

Berg, Christian, *Theologie im technologischen Zeitalter. Das Werk Ian Barbours als Beitrag zur Verhältnisbestimmung von Theologie zur Naturwissenschaft und Technik,* Stuttgart 2002.

– 'Barbour's Way(s) of Relating Science and Theology', in: Robert J. Russell (ed.), *Fifty Years in Science and Religion. Ian G. Barbour and His Legacy*, Aldershot 2004, 61-75.

Berger, Peter L., 'Protestantism and the Quest for Certainty', *The Christian Century* 115 (1998), 782-796.

– (ed.), *The De-Secularization of the World*, Grand Rapids 1999.

Berkhof, Hendrikus, 'God: voorwerp van wetenschap? II', in: idem, *Bruggen en bruggenhoofden*, Nijkerk 1981.

– *Introduction to the Study of Dogmatics*, Grand Rapids 1985 (or. Dutch edition: Kampen 1982).

– *Christian Faith. An Introduction to the Study of the Faith*, Grand Rapids 2002 (tr. of the fifth revised Dutch edition: Nijkerk 1985).

Beukel, A. van den, *More Things in Heaven and Earth. God and the Scientists*, London 1991 (original Dutch edition: Baarn 1990).

– *Waarom ik blijf. Gedachten over geloof, theologie en wetenschap*, Baarn 2003.

Bhaskar, Roy, *A Realist Theory of Science*, London 1997.

Biology and Gender Study Group, 'The Importance of Feminist Critiques for Contemporary Cell Biology', in: Nancy Tuana (ed.), *Feminism and Science*, Bloomington 1989, 172-187.

Blumenberg, Hans, *Die Legitimität der Neuzeit*, Frankfurt a.M. 1966.

Bonhoeffer, Dietrich, *Berlin 1932-1933* (Dietrich Bonhoeffer Werke, Volume 12), Gütersloh 1997.

Braithwaite, Richard B., 'An Empiricist's View of the Nature of Religious Belief', in: R. B. Braithwaite (ed.), *The Philosophy of Religion*, Oxford 1971, 72-91 (or. edition of the article: 1955).

Brink, Gijsbert van den, *Almighty God. The Doctrine of Divine Omnipotence*, Kampen 1996[2] (1993).

– 'Cultural Plausibility and Christian Theism', in: Henri A. Krop et al. (eds.), *Post-Theism. Reframing the Judaeo-Christian Tradition*, Leuven 2000, 265-275.

– *Oriëntatie in de filosofie*, Zoetermeer 2007[3] (2000).

– *Een publieke zaak. Theologie tussen geloof en wetenschap*, Zoetermeer 2004.

– 'Geloof en wetenschap: op zoek naar consonantie', in: Gijsbert van den Brink, Cees Dekker & René van Woudenberg (eds.), *Omhoog kijken in platland. Over geloven in de wetenschap*, Kampen 2007, 50-75.

– 'How to Speak with Intellectual and Theological Decency on the Resurrection of Christ? A Comparison of Swinburne and Wright', *Scottish Journal of Theology* 61 (2008), 408-419.

– *Als een schoon boec. Achtergrond, receptie en relevantie van artikel 2 van de Nederlandse Geloofsbelijdenis*, Leiden 2008.

Brooke, John Hedley, *Science and Religion. Some Historical Perspectives*, Cambridge 1991.

Bruce, James, *Travels to Discover the Sources of the Nile* V, Edinburgh 1790.

Brümmer, Vincent, *Theology and Philosophical Inquiry. An Introduction*, London 1981.

– *The Model of Love. A Study in Philosophical Theology*, Cambridge 1993.

– 'A Dialogue of Language Games', in: id. (ed.), *Interpreting the Universe as Creation*, Kampen 1991, 1-17.

– *Brümmer on Meaning and the Christian Faith. Collected Writings of Vincent Brümmer* (Ashgate Contemporary Thinkers on Religion), Aldershot 2006.

Bultmann, Rudolf, *The New Testament and Mythology, and Other Basic Writings*, Philadelphia 1984.

Buren, Paul M. van, *The Secular Meaning of the Gospel*, London 1963.

– *A Christian Theology of the People Israel* (A Theology of the Jewish-Christian Reality, Part 2), New York 1983.

Burke, Mary Anne & Stephen A Matlin (eds.), *Monitoring Financial Flows for Health Research 2008*, Geneva 2008.

Busch, Eberhard, *The Great Passion. An Introduction to Karl Barth's Theology*, Grand Rapids 2004 (or. German edition: Gütersloh 1998).

Calvin, John, *Institutes of the Christian Religion*, 2 vols., tr. by Ford Lewis Battles, Philadelphia 1960.

Cantor, Geoffrey & Chris Kenny, 'Barbour's Four-fold Way: Problems with His Taxonomy of Science-Religion Relationships', *Zygon* 36 (2001), 765-781.

Caputo, John D. & M.J. Scanlon (eds.), *God, the Gift, and Postmodernism*, Bloomington 1999.

Carnap, Rudolf, 'Testability and Meaning', *Philosophy of Science 3* (1936), 419-471.

– *Meaning and Necessity. A Study in Semantics and Modal Logic*, Chicago 1956.

Carson, D. A., *The Gagging of God. Christianity Confronts Pluralism*, Leicester 1996.

Casimir, Hendrik B.G., *Haphazard Reality. Half a Century of Science*, New York 1983.

Clayton, Philip, *Explanation from Physics to Theology*, London 1989.

– (ed.), *All That Is. A Naturalistic Faith for the 21st Century*, Minneapolis 2007.

Clayton, Philip & Steven Knapp, 'Is Holistic Justification Enough?', in: W. Mark Richardson & W. J. Wildman (eds.), *Religion & Science. History, Method, Dialogue*, New York 1996, 161-167.

Cliteur, Paul B., et al. (eds.), *Overtuigend bewijs. Over het wetenschappelijke van de niet-exacte wetenschappen*, Amsterdam 1994.

– *De neutrale staat, het bijzonder onderwijs en de multiculturele samenleving*, Leiden 2004.

Clouser, Roy, 'A Sketch of Dooyeweerd's Philosophy of Science', in: Jitse van der Meer (ed.), *Facets of Faith and Science*, Lanham 1996, Vol. 2, 81-98.

Cobb, John B. & David Ray Griffin, *Process Theology. An Introductory Exposition*, Philadelphia 1976.

Code, Lorraine, *What Can She Know? Feminist Theory and the Construction of Knowledge*, Ithaca 1991.

Corey, Michael A., 'Anthropic Principle', in: Van Huyssteen (ed.), *Encyclopedia of Science and Religion*, 13-18.

Craig, William Lane, 'Theism and Physical Cosmology', in: Philip L. Quinn & Charles Taliaferro (eds.), *A Companion to Philosophy of Religion*, Oxford 1997, 419-425.

Crombie, I. M., 'Theology and Falsification', in: A. Flew & A. MacIntyre, *New Essays in Philosophical Theology*, London 1955, 109-130.

Crowder, Colin (ed.), *God and Reality. Essays on Christian Non-Realism*, London 1997.

Curd, Martin & J.A. Cover, *Philosophy of Science. The Central Issues*, New York 1998.

Davies, Paul, *The Mind of God. Science and the Search for Ultimate Meaning*, London 1992.

Dawkins, Richard, *The Blind Watchmaker*, New York 1987.

D'Costa, Gavin, *Theology in the Public Square*, Oxford 2005.

Dear, Peter, *Revolutionizing the Sciences. European Knowledge and Its Ambitions, 1500-1700*, Princeton 2001.

Diem, Hermann, *Theologie als kirchliche Wissenschaft*, 3 Volumes, Munich 1951-1963.

Dijksterhuis, Eduard Jan, *The Mechanization of the World Picture. Pythagoras to Newton*, Princeton 1986 (original English tr. Oxford 1961, original Dutch ed. Amsterdam 1950).

Dilworth, Craig, *Scientific Progress. A Study Concerning the Nature of the Relation Between Successive Scientific Theories*, fourth edition, Berlin 2008 (or. ed. Dordrecht 1981).

Dippel, C.J. & J.M. de Jong (eds.), *Geloof en natuurwetenschap. Studies over de verhouding van christelijk geloof en moderne natuurwetenschap*, 2 Vols., The Hague 1965-1967.

Dooremalen, Hans, Herman de Regt & Maurice Schouten, *Exploring Humans: An Introduction to the Philosophy of the Social Sciences*, Amsterdam 2007.

Dooyeweerd, Herman, *A New Critique of Theoretical Thought*, 4 Volumes, Amsterdam 1953-1958.

Draper, John William, *History of the Conflict between Religion and Science*, New York 1875.

Drees, Willem B., *Religion, Science and Naturalism*, Cambridge 1996.

- *Over wijsbegeerte van de godsdienst en haar object: Twee formules*, Leiden 2002.
- *Creation: From Nothing Until Now*, London 2002.
- Review of *A Scientific Theology*, Volume 1: Nature, in: *Ars Disputandi* [http://www.arsdisputandi.org] 2 (2002), 3 August.
- '„Religion and Science" without Symmetry, Plausibility, and Harmony', *Theology and Science* 1 (2003), 113-128.

– Review of *Een publieke zaak*, in: *Nederlands Theologisch Tijdschrift* 60 (2006), 355-356.

Duhem, Pierre M.M., *The Aim and Structure of Physical Theory*, Princeton 1991 (original English tr. 1954; original French ed.: Paris 1906).

Farley, Edward, *Theologia. The Fragmentation and Unity of Theological Education*, Philadelphia 1983.

Farrar, Frederic William, *History of Interpretation*, London 1886.

Feyerabend, Paul K., *Against Method. Outline of an Anarchistic Theory of Knowledge*, London 1975.

Flew, Antony, 'Theology and Falsification', in: Flew & MacIntyre (eds.), *New Essays*, 96-99.

– (ed.), *A Dictionary of Philosophy*, London 1983.

Flew, Antony & Gary Habermas, 'My Pilgrimage from Atheism to Theism. A Discussion between Antony Flew and Gary Habermas', *Philosophia Christi* 6 (2004), 197-211.

Flew, Antony & Alasdair MacIntyre (eds.), *New Essays in Philosophical Theology*, London 1955.

Ford, David F. (ed.), *The Modern Theologians*, 2nd edition, Oxford 1997.

– *Theology. A Very Short Introduction*, Oxford 1999.

Fraassen, Bas C. van, *The Scientific Image*, Oxford 1980.

– *The Empirical Stance*, New Haven 2002.

Fuller, Steve, *Kuhn vs. Popper. The Struggle for the Soul of Science*, Cambridge 2003.

Gay, Peter. *The Enlightenment. A Comprehensive Anthology*, New York 1973.

Gergen, Kenneth R., *Realities and Relationships. Soundings in Social Constructionism*, Cambridge 1994.

Geurts, J.P.M., *Feit en theorie. Kernproblemen in de wetenschapsleer*, Assen 1984^3 (1975).

Giddens, Anthony, *New Rules of Sociological Method. A Positive Critique of Interpretative Sociologies*, London 1976.

Global Forum for Health Research, *10/90 Report on Health Research 2003-2004*, Geneva 2004.

Gottlieb, Roger S., *This Sacred Earth. Religion, Nature and Environment*, New York 1996.

Gootjes, Nicolaas H., *The Belgic Confession. Its History and Sources*, Grand Rapids 2007.

Gould, Stephen Jay, *Rocks of Ages. Science and Religion in the Fullness of Life*, New York 1999.

Groot, A. D. de, *Methodology. Foundations of Inference and Research in the Behavioral Sciences*, The Hague 1969 (or. Dutch edition: The Hague 1961).

Groot, A. D. de & H. Visser, *Het forumwaarmerk van wetenschap*, Amsterdam 2003.

Habermas, Jürgen, 'Knowledge and Human Interests: A General Perspective', in: id., *Knowledge and Human Interests*, 2nd ed., London 1978 (or. German edition 'Erkenntnis und Interesse. Frankfurter Antrittsvorlesung vom 28.6.1965', in: *Merkur* 213, December 1965, 1139-1153).

Hailer, M., *Theologie als Weisheit*, Neukirchen 1997.

Hall, Alfred Rupert, *The Revolution in Science, 1500-1750*, London 1983.

Hanson, Norwood Russell, *Patterns of Discovery. An Inquiry into the Conceptual Foundations of Science*, Cambridge 1958.

Harding, Sandra, *The Science Question in Feminism*, Ithaca 1986.

Hardy, Daniel & David Ford, *Praising and Knowing God*, Philadelphia 1985.

Hark, Michel ter, *Popper, Otto Selz and the Rise of Evolutionary Epistemology*, Cambridge 2004.

Hawking, Stephen, *A Brief History of Time*, London 1988.

Heisenberg, Werner, *Physics and Beyond. Encounters and Conversations*, New York 1971 (tr. of *Der Teil und das Ganze. Gespräche im Umkreis der Atomphysik*, Munich 1969).

Helm, Paul, *Faith with Reason*, Oxford 2000.

Hempel, Carl G., *Aspects of Scientific Explanation, and Other Essays in the Philosophy of Science*, New York 1965.

Hempel, Carl G. & Paul Oppenheim, 'Studies in the Logic of Explanation', *Philosophy of Science 15* (1948), 137-175.

Henle, Paul (ed.), *Language, Thought and Culture*, Ann Arbor 1965.

Hesse, Mary B., *Revolutions and Reconstructions in the Philosophy of Science*, Brighton 1980.

Hick, John (ed.), *The Existence of God*, London 1964.

Hitchcock, Christopher (ed.), *Contemporary Debates in the Philosophy of Science*, Oxford 2004.

Holten, Wilko van, *Explanation within the Bounds of Religion*, Frankfurt 2003.

Hooykaas, Reijer, 'Science and Reformation', *The Journal of World History* 3 (1956), 136-138.

– *Religion and the Rise of Modern Science*, Grand Rapids 1972.

Horsten, Leon, Igor Douven & Erik Weber, *Wetenschapsfilosofie*, Assen 2007.

Houtepen, Anton, *God: An Open Question,* New York 2002.

Howell, Kenneth J., *God's Two Books. Copernican Cosmology and Biblical Interpretation in Early Modern Science*, Notre Dame 2002.

'Hoyle on Evolution', *Nature* 294, no. 5837 (12 November 1981), 105.

Hubbeling, H.G., *Principles of the Philosophy of Religion*, Assen/Maastricht 1987.

Huiser, Pieter, *Models, Theories, Narratives. Conditions for the Justification of a Religious Realism*, Amsterdam 1997.

Husserl, Edmund, *Logical Investigations* I, London 1970 (original German ed.: Halle 1900).

Huyssteen, J. Wentzel van, *Theology and the Justification of Faith*, Grand Rapids 1989.

– *Essays in Postfoundationalist Theology*, Grand Rapids 1997.

– *The Shaping of Rationality.Toward Interdisciplinarity in Theology and Science*, Cambridge 1999.

– *Alone in the World? Human Uniqueness in Science and Religion*, Grand Rapids 2006.

– (ed.), *Encyclopaedia of Science and Religion*, New York 2003.

Immink, F. Gerrit, *Faith. A Practical Theological Reconstruction*, Grand Rapids 2005 (or. Dutch edition: Zoetermeer 2003).

Janicaud, Dominique, et al., *Phenomenology and the Theological Turn. The French Debate*, New York 2000.

Joest, Wilfred, *Fundamentaltheologie: Theologische Grundlagen- und Methodenprobleme*, Stuttgart 1974.

Jongeneel, Jan A.B., 'The Missiology of Gisbertus Voetius: The First Comprehensive Protestant Theology of Missions', *Calvin Theological Journal* 26 (1991), 47-79.

Keller, Evelyn Fox, *A Feeling for the Organism. The Life and Work of Barbara McClintock*, San Francisco 1983.

– *Reflections on Gender and Science*, New Haven 1985.

Keulen, Dirk van, 'The Internal Tension in Kuyper's Doctrine of Organic Inspiration of Scripture', in: Cornelis van der Kooi & Jan de Bruijn (eds.), *Kuyper Reconsidered. Aspects of his Life and Work*, Amsterdam 1999, 123-130.

Kim, Jaegwon, *Philosophy of Mind*, Oxford 1996.

Koningsveld, Herman, *Het verschijnsel wetenschap*, Meppel/Amsterdam 1980 (or. ed. 1976; new revised ed. Amsterdam 2006).

Kuhn, Thomas S., *The Structure of Scientific Revolutions*, 2nd ed., Chicago 1970 (orig. ed. Chicago 1962).

Kuipers, Theo A.F., *Structures in Science. Heuristic Patterns Based on Cognitive Structures*, Dordrecht 2001

– 'Poppers filosofie van de natuurwetenschappen', *Wijsgerig Perspectief* 42.2 (2002), 17-31.

Kuitert, H.M., *Filosofie van de theologie*, Leiden 1988.
– *Jesus: The Legacy of Christianity*, London 2000.
Küng, Hans & David Tracy (eds.), *Paradigm Change in Theology*, Edinburgh 1989.
Kuyper, Abraham, *To Be Near unto God*, Vancouver BC 2005 (or. edition: Grand Rapids 1918; or. Dutch edition: Kampen 1908).
Labooy, Guus, *Freedom and Dispositions – Two Main Concepts in Theology and Biological Psychiatry. A Systematic Analysis*, Frankfurt a.M. 2000.
Lakatos, Imre, *Philosophical Papers*, Volume 1, Cambridge 1978.
Lakatos, Imre & Paul K. Feyerabend, *For and Against Method. Including Lakatos's Lectures on Scientific Method and the Lakatos-Feyerabend Correspondence*, Chicago 1999.
Lakatos, Imre & Alan Musgrave (eds.), *Criticism and the Growth of Knowledge*, Cambridge 1970.
Laudan, Larry L., *Progress and its Problems. Towards a Theory of Scientific Growth*, Berkeley 1977.
Leezenberg, Michiel & Gerard de Vries, *Wetenschapsfilosofie voor geesteswetenschappen*, Amsterdam 2001.
Livingstone, David N., *Darwin's Forgotten Defenders. The Encounter between Evangelical Theology and Evolutionary Thought*, Edinburgh 1987.
Lloyd, Genevieve, *The Man of Reason. „Male" and „Female" in Western Philosophy*, London 1984.
Löwith, Karl, *Meaning in History. The Theological Implications of the Philosophy of History*, Chicago 1949.
Longino, Helen E., *Science as Social Knowledge. Values and Objectivity in Scientific Inquiry*, Princeton 1990.
Loonstra, Bert, *God schrijft geschiedenis. Disputaties over de Eeuwige*, Zoetermeer 2003.
MacIntyre, Alisdair, *After Virtue. A Study in Moral Theory*, London 1981.
– *Whose Justice? Which Rationality?*, Notre Dame 1988.
– *Three Rival Versions of Moral Inquiry*, Notre Dame 1990.
MacMullin, Ernan, 'How Should Cosmology Relate to Theology?', in: Arthur Peacocke (ed.), *The Sciences and Theology in the Twentieth Century*, Notre Dame 1981, 17-57.
– (ed.), *The Church and Galileo*, Notre Dame 2005.
Mannheim, Karl, 'Die Bedeutung der Konkurrenz im Gebiete des Geistigen' (Vortrag auf dem Züricher Soziologentag 1928), in: Volker Meja & Nico Stehr (eds.), *Der Streit um die Wissenssoziologie* 1, Frankfurt 1982, 325-370.

Marion, Jean-Luc, *Being Given. Toward a Phenomenology of Givenness*, Stanford 2002.

Markus, Arjan, *Beyond Finitude. God's Transcendence and the Meaning of Life*, Frankfurt 2004.

Marshall, I. Howard, *1 Peter*, Leicester 1991.

Mathewes, Charles, 'Political Theology After the End of History', paper presented at the Annual Conference of the Society for the Study of Theology, Durham (UK), April 2008.

Mavrodes, George, *Belief in God. A Study in the Epistemology of Religion*, New York 1970.

McGrath, Alister E., *The Foundations of Dialogue in Science and Religion*, Oxford 1998.

– *Science and Religion. An Introduction*, Oxford 1999.

– *A Scientific Theology*, 3 vols., Edinburgh 2001-2003.

– *The Science of God. An Introduction to Scientific Theology*, Edinburgh 2004.

Mitchell, Basil, *The Justification of Religious Belief*, London 1973.

Molendijk, Arie L., *Aus dem Dunklen ins Helle. Wissenschaft und Theologie im Denken von Heinrich Scholz*, Amsterdam 1991.

Moore, James R., *The Post-Darwinian Controversies*, Cambridge 1979.

Murphy, Nancey, *Theology in the Age of Scientific Reasoning,* Ithaca 1990.

Murphy, Nancey, & James McClendon, 'Distinguishing Modern and Postmodern Epistemologies', *Modern Theology* 5 (1989), 191-214.

Nelson, Lynn H., *Who Knows? From Quine to a Feminist Empiricism*, Philadelphia 1990.

Neurath, Otto, *Empiricism and Sociology*, Dordrecht 1973.

Newberg, Andrew, Eugene d'Aquili & Vincent Rause, *Why God Won't Go Away. Brain Science and the Biology of Belief*, New York 2001.

Newbigin, Lesslie, *The Gospel in a Pluralist Society*, London 1989.

– *Truth to Tell. The Gospel as Public Truth*, London 1991.

– *Proper Confidence. Faith, Doubt and Certainty in Christian Discipleship*, Grand Rapids 1995.

Nielsen, Kai, 'Wittgensteinian Fideism', *Philosophy* 42 (1967), 191-209.

Noordmans, Oepke, *Herschepping*, Zeist 1934.

Okasha, Samir, *Philosophy of Science. A Very Short Introduction*, Oxford 2002.

Okruhlik, Kathleen, 'Gender and the Biological Sciences', *Canadian Journal of Philosophy* (Suppl. Volume) 20 (1994), 21-42.

Ouweneel, Willem J., *Christian Doctrine. Old Truths in New Perspectives* 1, Amsterdam 1995.

Padgett, Alan G., *Science and the Study of God. A Mutuality Model for Theology and Science*, Grand Rapids 2003.

Pannenberg, Wolfhart, *Theology and the Philosophy of Science*, Philadelphia (Pa.) 1976 (or. German edition: Frankfurt am Main 1973).

– *Systematic Theology* Vol. 1, Grand Rapids 1991 (or. German edition: Göttingen 1988).

– *Theologie und Philosophie. Ihr Verhältnis im Lichte ihrer gemeinsamen Geschichte*, Göttingen 1996.

– *The Historicity of Nature*, ed. by Niels Henrik Gregersen, West Conshohocken 2008.

Parker, Michael G., & Thomas M. Schmidt (eds.), *Scientific Explanation and Religious Belief. Science and Religion in Philosophical and Public Discourse*, Tübingen 2005.

Peacocke, Arthur R., *Theology for a Scientific Age*, London 1993.

– *Paths from Science towards God*, Oxford 2001.

Pelikan, Jaroslav, *Christianity and Classical Culture. The Metamorphosis of Natural Theology in the Christian Encounter with Hellenism*, New Haven 1993.

– *Credo. Historical and Theological Guide to Creeds and Confessions of Faith in the Christian Tradition*, New Haven 2003

Peters, Ted (ed.), *Science and Theology. The New Consonance*, Boulder 1998.

Peterson, Michael, et al., *Reason & Religious Belief. An Introduction to the Philosophy of Religion*, 2nd edition, New York 1998 (or. ed. 1990).

Philipse, Herman, 'The Incompatibility of Science and Religion: An Argument for Atheism', in: William Desmond, John Steffen & Koen Decoster (eds.), *Beyond Conflict and Reduction. Between Philosophy, Science, and Religion*, Leuven 2001, 117-134.

Phillips, D.Z., *Faith after Foundationalism*, London 1988.

Pinker, Steven, *The Language Instinct. The New Science of Language and Mind*, New York 1994.

Plantinga, Alvin, 'Methodological Naturalism?', in: Jitse M. van der Meer (ed.), *Facets of Faith and Science*, Lanham 1996, part I, 177-221.

– *Warranted Christian Belief*, Oxford 2000.

Plantinga, Alvin, & Nicholas Wolterstorff (eds.), *Faith and Rationality. Reason and Belief in God*, Notre Dame 1983.

Polanyi, Michael, *Personal Knowledge. Towards a Post-Critical Philosophy*, London 1958.

Polkinghorne, John, *The Way the World Is*, London 1983.

– *Belief in God in an Age of Science*, New Haven 1988.

– *Science and Providence*, London 1989.

Popper, Karl R., *The Logic of Scientific Discovery*, New York 1965 (or. edition: Vienna 1935).

– *The Open Society and Its Enemies*, 2 Volumes, London 1945.

– *The Poverty of Historicism*, London 2002 (or. edition: London 1957).

– *Conjectures and Refutations. The Growth of Scientific Knowledge*, London 1963.

– *Unended Quest. An Intellectual Autobiography*, London 1976.

– *Realism and the Aim of Science. From the Postscripts to The Logic of Scientific Discovery*, London 1983.

Quine, Willard Van Orman, *From a Logical Point of View,* Cambridge (Mass.) 1953.

Quine, Willard Van Orman & J. S. Ullian, *The Web of Belief*, New York 1979.

Ratzsch, Del, *Science & Its Limits. The Natural Sciences in Christian Perspective*, Downers Grover 2000.

Reichenbach, Hans, *Experience and Prediction. An Analysis of the Foundations and the Structure of Knowledge*, Chicago 1938.

Reynolds, Jack, *Understanding Existentialism*, Chesham 2006.

Richardson, W. Mark & Wesley J. Wildman (eds.), *Religion & Science. History, Method, Dialogue*, New York 1996.

Rorty, Richard, *Philosophy and the Mirror of Nature*, Princeton 1980.

Rosen, E., 'Calvin's Attitude towards Copernicus', *Journal of the History of Ideas* 21 (1960), 431-441.

Ruetsche, Laura, 'Interpreting Quantum Mechanics', in: Peter Machamer & Michael Silberstein (eds.), *The Blackwell Guide to the Philosophy of Science*, Oxford 2002, 199-226.

Ruler, A.A. van, *Verzameld werk* 1 (De aard van de theologie), edited by Dirk van Keulen, Zoetermeer 2007.

Russell, Bertrand, *The Impact of Science on Society*, London 1952.

– *A History of Western Philosophy and Its Connections with Political and Social Circumstances from the Earliest Times to the Present Day*, 2nd ed., London 1961 (or. ed. 1946).

Russell, Robert John, et al. (eds.), *Chaos and Complexity. Scientific Perspectives on Divine Action*, Vatican City State 1995.

– (eds.), *Quantum Mechanics. Scientific Perspectives on Divine Action*, Vatican City State 2001.

Ryle, Gilbert, *Dilemmas. The Tarner Lectures 1953*, Cambridge 1954.

Salmon, Wesley C., 'Four Decades of Scientific Explanation', in: Philip Kitcher and Wesley C. Salmon (eds.), *Scientific Explanation* (Minnesota Studies in the Philosophy of Science 13), Minneapolis 1989, 3-219.

Sanders, Andy F., *Michael Polanyi's Post-Critical Philosophy. A Reconstruction of Some Aspects of 'Tacit Knowing'*, Amsterdam 1988.

Sarot, Marcel, *De goddeloosheid van de wetenschap*, Zoetermeer 2006.

Saunders, N.T., *Divine Action and Modern Science*, Cambridge 2002.

Sauter, Gerhard (ed.), *Theologie als Wissenschaft. Aufsätze und Thesen*, Munich 1971.

– (ed.), *Wissenschaftstheoretische Kritik der Theologie. Die Theologie und die neuere wissenschaftstheoretische Diskussion*, Munich 1973.

Schilpp, Paul A., *The Philosophy of Rudolf Carnap*, La Salle (Ill.) 1963.

Segal, Robert, 'In Defence of Reductionism', in: R. T. McCutcheon (ed.), *The Insider/Outsider Problem in the Study of Religion*, London 1999, 139-163.

Sehon, Scott, *Teleological Realism. Mind, Agency, and Explanation*, Cambridge (MA) 2005.

Smedes, Taede A., *Chaos, Complexity, and God. Divine Action and Scientism*, Leuven 2004.

Spinks, D. Christopher, *The Bible and the Crisis of Meaning. Debates on the Theological Interpretation of Scripture*, London n.y. [2007].

Stenmark, Mikael, *Rationality in Science, Religion, and Everyday Life*, Notre Dame 1995.

– *Scientism. Science, Ethics and Religion*, Aldershot 2001.

Stewart, Robert B. (ed.), *Intelligent Design. William A. Dembski & Michael Ruse in Dialogue*, Minneapolis 2007.

Stoker, Wessel, *Is Faith Rational? A Hermeneutical-Phenomenological Approach*, Leuven 2006 (or. Dutch edition: Zoetermeer 2004).

Storrar, William F., & Andrew R. Morton (eds.), *Public Theology for the 21st Century*, Edinburgh 2004.

Strevens, Michael, *Depth. An Account of Scientific Explanation*, Cambridge (MA) 2008.

Suppe, Frederick (ed.), *The Structure of Scientific Theories*, 2nd edition, Urbana 1977 (or. ed. 1974).

Swinburne, Richard, *The Existence of God*, revised edition, Oxford 1991.

– *Is there a God?* Oxford 1996.

– *The Resurrection of God Incarnate*, Oxford 2003.

Thiemann, Ronald F., *Revelation and Theology. The Gospel as Narrated Promise*, Notre Dame 1985.

Torrance, Thomas F., *Theological Science*, London 1971.

– *Divine and Contingent Order*, Oxford 1981.
Toulmin, Stephen E., *Foresight and Understanding. An Inquiry into the Aims of Science*, London 1961.
Tracy, David, *The Analogical Imagination. Christian Theology and the Culture of Pluralism*, New York 1981.
– 'Defending the Public Character of Theology', *The Christian Century* 98 (1981), 350-356.
– 'Theology and the Many Faces of Postmodernity', *Theology Today* 51 (1994), 104-114.
Trigg, Roger, *Rationality and Religion*, Oxford 1998.
– *Religion in the Public Life. Must Faith be Privatized?*, Oxford 2007.
Vanderjagt, Arjo & Klaas van Berkel (eds.), *The Book of Nature in Antiquity and the Middle Ages*, Leuven 2005.
Vanhoozer, Kevin (ed.), *Dictionary for Theological Interpretation of the Bible*, Grand Rapids 2005.
Voetius, Gisbertus, *De pietate cum scientia conjungenda* (inaugural address, Utrecht 1634).
Vos, Antonie, et al., *John Duns Scotus, Contingency and Freedom*, introduction, translation and commentary, Dordrecht 1994.
Wainwright, Geoffrey, *Doxology. The Praise of God in Worship, Doctrine and Life. A Systematic Theology*, New York 1980.
Weizsäcker, C.F. von, *The World View of Physics*, London 1952 (or. German ed.: Leipzig 1943).
– *Die Tragweite der Wissenschaft*, 6th enlarged ed., Stuttgart 1990 (or. ed. 1964).
Welch, Claude, 'Dispelling Some Myths about the Split between Theology and the Sciences in the Nineteenth Century', in: Richardson & Wildman (eds.), *Religion & Science*, New York 1996, 29-40.
Wenz, Gunter, *Wolfhart Pannenbergs Systematische Theologie*, Göttingen 2003.
White, Andrew Dickson, *The Warfare of Science*, New York 1876.
– *A History of the Warfare of Science with Theology in Christendom*, New York 1896.
Whitehead, Alfred N., *Process and Reality. An Essay in Cosmology*, corrected edition, New York 1978 (or. edition 1929).
Wiggershaus, Rolf, *The Frankfurt School. Its History, Theories, and Political Significance*, Cambridge (MA) 1994 (or. German edition: Munich 1986).
Wilde, Rein de, 'Verklaren in de sociale wetenschappen', in: Michiel Korthals (ed.), *Wetenschapsleer. Filosofisch en maatschappelijk perspectief op de natuur- en sociaal-culturele wetenschappen*, Meppel 1989, 94-111.

Wildman, Wesley J., 'The Divine Action Project, 1998-2003', *Theology and Science* 2 (2004), 31-75.

Williams, Rowan, *On Christian Theology*, London 1999.

Wilson, Edward O., *Consilience. The Unity of Knowledge*, New York 1998.

Wittgenstein, Ludwig, *Philosophical Investigations*, Oxford 1953.

– *Lectures and Conversations on Aesthetics, Psychology and Religious Belief*, ed. C. Barrett, Oxford 1966.

Wolterstorff, Nicholas, *Reason within the Bounds of Religion*, Grand Rapids 1976.

Woudenberg, René van, *Gelovend denken. Inleiding tot een christelijke filosofie*, Kampen 1992.

– *Toeval en ontwerp in de wereld. Apologetische analyses*, Budel 2003.

– 'Aspects and Functions of Individual Things', *Philosophia Reformata 68* (2003), 1-13.

Wright, N.T., *The Resurrection of the Son of God: Christian Origins and the Question of God*, London 2003.

Young, Davis A., *John Calvin and the Natural World*, Lanham 2007.

Contributions to Philosophical Theology

Edited by Gijsbert van den Brink, Vincent Brümmer and Marcel Sarot

Vol. 1 Gijsbert van den Brink / Marcel Sarot (eds.): Understanding the Attributes of God. 1999.

Vol. 2 Marcel Sarot / Gijsbert van den Brink (eds.): Identity and Change in the Christian Tradition. 1999.

Vol. 3 Marcel Sarot: Living a Good Life of Evil. 1999.

Vol. 4 William Hasker / David Basinger / Eef Dekker (eds.): Middle Knowledge. Theory and Applications. 2000.

Vol. 5 Wybren de Jong: Identities of Christian Traditions. An Alternative for Essentialism. 2000.

Vol. 6 Eeva Martikainen (ed.): Infinity, Causality and Determinism. Cosmological Enterprises and their Preconditions. 2002.

Vol. 7 Gerrit Brand: Speaking of a Fabulous Ghost. In Search of Theological Criteria, with Special Reference to the Debate on Salvation in African Christian Theology. 2002.

Vol. 8 Guus Labooy: Freedom and Dispositions. Two Main Concepts in Theology and Biological Psychiatry, a systematic Analysis. 2002.

Vol. 9 Wilko van Holten: Explanation within the Bounds of Religion. 2003.

Vol. 10 Santiago Sia: Religion, Reason and God. Essays in the Philosophies of Charles Hartshorne and A.N. Whitehead. 2004.

Vol. 11 Arjan Markus: Beyond Finitude. God's Transcendence and the Meaning of Life. 2004.

Vol. 12 Gijsbert van den Brink: Philosophy of Science for Theologians. An Introduction. 2009.

www.peterlang.de

www.ingramcontent.com/pod-product-compliance
Ingram Content Group UK Ltd.
Pitfield, Milton Keynes, MK11 3LW, UK
UKHW020834290926
15470UKWH00004B/1348